THE
ELECTRICAL SENSITIVITY
HANDBOOK

How Electromagnetic Fields (EMFs)
Are Making People Sick

Lucinda Grant

Weldon Publishing
Post Office Box 4146
Prescott, Arizona 86302

ISBN: 0-9635407-2-6

Library of Congress Catalog Card Number: 95-60142

Printed in the United States of America

First Edition

Chemical and electromagnetic pollution — the residue of modern industrialization — show their price in the global escalation of cancer and environmental illnesses.

The American Cancer Society's publication <u>Cancer Facts & Figures - 1993</u> predicts that one out of every three Americans now living will have cancer. Currently, one out of every five deaths is from cancer. [26]

Note: Numbered references throughout the text indicate sources for further study.
(See Bibliography, page 89.)

This book is dedicated to

electrically sensitive patients worldwide.

Outstanding among

those patients who helped

with ideas for this book are those in

The Association for the Electrically and VDT Injured,

Post Office Box 15126,

10465 Stockholm, Sweden.

May we all find peace, health and happiness — everyone.

TABLE OF CONTENTS

Disclaimer

The author is an electrically sensitive patient, but not a medical doctor or scientist.

For your protection, seek respected medical or other appropriate advice prior to making any changes in your current program.

Electrical sensitivity is a new medical area. As more research and surveys are completed, facts about electrical sensitivity will become more clear and may change from the initial impressions presented in this book.

This document is intended to provide general information to the public about electrical sensitivity and electromagnetic fields to facilitate discussion and legislation in this area.

The author and the publisher are not responsible for any damage or loss of any kind occurring from information in this book, including errors 'and omissions. The information contained herein is only meant to suggest possibilities and cause the reader to seek proper reliable advice as pertains to his/her individual situation.

Electrical Sensitivity (ES) is Important for *Everyone*:

- ES shows that the body can be sensitive to very low electromagnetic field (EMF) levels, below those levels now considered safe.

- Cancer is primarily discussed as the main EMF health risk. When ES patients show an EMF reaction at 15-20 feet from a refrigerator, does that mean cancer could be promoted (or caused) at that distance also?

- Electrically sensitive patients have shown that external EMFs are biologically active and can interfere with the electricity of life processes, particularly the nervous system. EMFs may be neurotoxic.

- Electrical sensitivity can explain health problems many people are presently having for which they currently have no answers and no long-term relief.

- ES patient reactions represent the subtle nature of life and the EMF forces that control and maintain it.

- ES pinpoints the need for environmental regulation of EMF pollution in all its forms.

Electrical Sensitivity Network

The need for a national electrical sensitivity network, a support and advocacy group, is expressed in this letter reproduced in part from Network News, October/November 1994 (See EMF Resources, page 74.):

"People with electromagnetic sensitivity (ES) disorder often find themselves disabled, misunderstood, and unrepresented. There seems to be no group organized on a national level specifically to coordinate the needs of people with ES in the United States. Germany, Sweden, and Denmark do have such associations.

With these thoughts in mind and as an ES patient myself, I hope to help develop a list of resources and a list of ES patients interested in starting a group focusing on ES needs. Group goals could entail idea sharing, emotional support, and being heard as a group for proper representation under the Americans with Disabilities Act, and with federal and state agencies working with the injured and/or disabled.

A survey of ES patients indicating symptoms, patient profile, and most severe electromagnetic triggers, and coping methods, could give weight to the proper recognition of ES as a widespread U.S. illness. A major question is "How many people here have ES?". In Sweden about 1,500 patients belong to their Association for the Electrically and VDT Injured.

People who develop symptoms which intensify near electrical appliances and other electromagnetic field sources find little consolation from general medical or scientific circles. Therefore, ES patients are frequently without proper documentation to pursue legal action, explain their symptoms to others, or retain/obtain suitable employment if affected in the workplace.

As a group, I hope we can make ES more understood and acknowledged as an illness/disability."

For more information about this group now forming, contact Weldon Publishing, Post Office Box 4146, Prescott, Arizona 86302.

Terminology

Electrical sensitivity (ES) is a modern illness of the technological age. A review of available literature reveals that ES has been known by the following terms:

Electric Hypersensitivity

Electrical Hypersensitivity

Electrical Oversensitivity

Electrical Sensitivity (ES)

Electromagnetic Field (EMF) Hypersensitivity

Electromagnetic Field Sensitivity

Electromagnetic Hypersensitivity

Electromagnetic Hypersensitivity Syndrome

Electromagnetic Sensitivity

Electrosensitivity

Electrosupersensitivity

* Screen Dermatitis

* "Terminal" Dermatitis

* Video Operator's Distress Syndrome (VODS)

The term *electrical sensitivity* seems to be generally understood and acceptable for United States and foreign communications. *Electrical sensitivity* also suits the term *chemical sensitivity*, another part of the environmental illness picture.

* Terms relating to electrical sensitivity symptoms specific to computer users.

What is Environmental Illness?

Environmental illness, also known as EI, means an illness occurring because of exposure to environmental toxins. Environmental illness is chemical sensitivity, electrical sensitivity, the Sick Building Syndrome, and now probably the Gulf War Syndrome.

Veterans from the Persian Gulf War, who are collectively ill from symptoms such as fatigue, headaches, nausea, breathing and memory problems, have called their illness Gulf War Syndrome. This syndrome is currently suspected of being related to toxic chemical exposure and, if so, makes the illness another form of chemical sensitivity.[79]

Multiple chemical sensitivity (MCS) occurs when someone is chemically poisoned by toxins in the environment and subsequently becomes ill, especially when exposed to ordinary chemicals in our modern world.[66] Once overexposed, they can no longer tolerate such chemicals as household detergents, garden pesticides, perfumes, natural gas emissions from gas stoves, etc. Molds, dust, smoke, and pollen may also increase their symptoms.[94]

Dealing with chemical sensitivity means adjusting your environment to minimize irritants, indoors and out. Moving to an area with cleaner air, low pollen levels, and no pesticide spraying is ideal for controlling your outdoor environment to help your health.

Cleaning your indoor environment of chemicals that could be adding to your chemical load of exposures is often more difficult. Organizations such as The National Center for Environmental Health Strategies can suggest suitable alternative products to reduce chemical exposures. (See Information Resources, page 72.)

However, the most difficult part of having an environmental illness is not being able to control your environment beyond your home. When you leave home for errands or work, you are exposed to various everyday sorts of things that previously you would not have noticed, but now make the symptoms worse. The most chemically ill feel they cannot safely leave their home due to chemicals they may encounter everywhere else.

Chemically sensitive patients sometimes also become electrically sensitive. Their illness symptoms intensify near power lines and electrical appliances, causing some people to become hermits in order to avoid the twentieth century's modern conveniences.

What is Electrical Sensitivity?

Electrical sensitivity (ES) is a form of environmental illness — a chronic illness triggered by exposure to electromagnetic fields. (See What Are Electromagnetic Fields (EMFs)?, page 61.)

Being electrically sensitive means having an illness that *noticeably* reacts or intensifies near electrical appliances, power lines, and/or other electromagnetic field (EMF) sources. ES includes *recurring* feelings of stress or illness when near these EMF sources, even if the person affected has no illness apparent when not exposed to EMFs. Any noticeable, recurring ill health that is triggered by an electromagnetic field, and that diminishes or disappears away from the EMF source, constitutes a case of electrical sensitivity.

Just as the chemically sensitive can become very sensitive to small amounts of chemicals, ES patients can become *hypersensitive* to EMF levels that normally would not be noticeable to the general public. However, the dividing line between "safe" and unsafe levels of EMF exposure is not presently clear for the public at large. (See The EMF Cancer Issue, page 62.)[2, 40, 41, 53, 69]

People sometimes become ill from an EMF source, such as having an electrical appliance (i.e., clock, cassette player, etc.) too close to their bed. Published accounts mention symptoms such as insomnia, nausea, headache, or other noticeable health problems from prolonged exposure to these EMFs.[39, 65]

In one United States case, a multiple chemical sensitivity patient developed nausea after meals in addition to his usual MCS symptoms. He thought about what he was doing differently and began to suspect that maybe his new health problem had something to do with his new VCR (video cassette recorder). He talked to friends about EMFs. One friend suggested he buy a gaussmeter to measure EMFs in his house. He did and subsequently measured EMF exposure in his house, including his appliances. A new cassette player on his bed's headboard gave a high reading, even when it was shut off. He moved the cassette player away from his bed. His nausea gradually decreased from that point on.[65]

Other similar instances were reported in a United States newsletter that interviewed German EMF testing technician, Wolfgang Maes. Three cases of people with problems sleeping were noted, all corrected by removing everything electrical from the bedroom. In addition, a child who had stomach pains, nervousness, insomnia, and headaches was able to sleep without headaches after the EMF wall wiring problems of his bedroom were solved. One day after the EMF was reduced, all of his health problems began to improve. A fifth case mentioned was a two-year-old with headaches and severe muscle cramps all over. She slept with a radio alarm clock on under her pillow to stay warm. When the parents took away the radio, the child gradually improved.[39]

In the cases cited, moving the offending EMF source(s) or correcting wall wiring problems produced a return to health or substantial improvement in health.

Were these people electrically sensitive? Although their health problems were not permanent, for a time they seemed to be electrically sensitive to a specific EMF source or location. If the EMF source had not been found and their EMF exposure reduced, they may have developed a more severe and permanent ES.

There are degrees of ES just as there are degrees of chemical sensitivity. Allergies are a mild form of chemical sensitivity; Sick Building Syndrome, a more severe form. Multiple chemical sensitivity is chemical sensitivity's most severe and chronic form.[78] The one-time, reversible form of ES from the cases noted above seem to be representative of a mild form of ES.

With chronic, long-term ES, the person becomes sensitive to an initiating EMF frequency, intensity, or duration. Later, the illness is further triggered when the person is exposed to other EMF sources also. This environmental illness process is called a *spreading phenomenon*.[94] A one-time ES experience should be considered a warning sign of possible chronic ES susceptibility.[14]

Electrical sensitivity has the great potential for being a missing link that uncovers explanations for many noticeable, recurring health problems people currently have no medical answers for and no long-term relief.

ES Statistics

How many people in the United States are estimated to be electrically sensitive (ES)?

No formal statistics are currently available.

As an estimate, we could assume that the ratio of ES patients to the general population is the same for both Sweden and the United States. In 1994, Sweden's Association for the Electrically and VDT Injured had about 1800 members. Further, Sweden's group estimated that in 1993 at least 10,000 Swedish citizens were ES.[14] Using Sweden's estimated range of ES people of between 1,800 - 10,000, the United States would have between 52,043 - 289,130.*

If mild forms of ES and other EMF-related illnesses are included, these figures would be significantly greater.

What percentage of chemically sensitive patients become ES?

Only an estimate is possible at this time.

The Environmental Health Center in Dallas has treated over 20,000 chemically sensitive patients at various stages of the illness. Also, they have treated over 500 ES patients.[94, 101] Some people had both illnesses and others only had one. If we assume that *all* of the ES patients were also chemical sensitivity patients, then as many as 2.5% of the chemically sensitive patients were ES too.

If only the most severe MCS patients were counted, the percentage of ES within that group would probably be more.

What Are the Symptoms of Electrical Sensitivity?

In general, environmental illnesses can produce nearly any symptom, depending

---

* 1992 Sweden population (estimate) = 8,602,000
  1990 United States population (estimate) = 248,709,873

upon the type of irritating exposure and the uniqueness of the individual exposed.[94] One exposure, whether chemical or electrical, can create symptoms of fatigue in one person and hyperactivity in another person. Symptoms of electrical sensitivity (ES) may manifest as any of a wide variety of health problems. However, symptoms are mainly skin-related or neurological.[12, 18] ES skin reactions tend to be a mild form of ES initially, but can intensify to intense burning levels with prolonged EMF exposure. Advanced ES skin reactions may be accompanied by neurological and other symptoms as well.

> Common ES symptoms when exposed to EMF sources are headache, nausea, fatigue, dizziness, tingling or prickling sensation on the skin, burning skin or eyes, difficulty concentrating, memory loss, muscle or joint pain, and heart rate fluctuations. Less common, but more severe reactions include paralysis, seizures, and unconsciousness.

Other symptoms may also be present; those listed prior appear to be representative of ES patients worldwide.[4, 14, 53, 69, 102] Several ES symptoms may occur together or only one. Because electromagnetic fields are almost everywhere in our modern society, even one chronic symptom can lead to restricted work/social opportunities and disability.

In the United States, a medical study of 100 ES patients at the Environmental Health Center, Dallas, Texas, showed many EMF reactions under controlled conditions: neurological, muscle/skeletal, heart, eye, ear, skin, tooth, respiratory, and gastrointestinal. The primary symptoms of these ES patients during the study was neurological — tingling, headache, sleepiness, dizziness, and unconsciousness.[102]

The Association for the Electrically and VDT Injured in Stockholm, Sweden list common ES symptoms for them as skin — burning, pricking, tingling; eyes — burning, dryness; ear, nose, throat — swollen membranes, dryness of nose and throat; neurological — memory problems, dizziness, lack of concentration, headache; heart — fluctuating heartbeat; tooth — pain; muscles/joints — pain; gastrointestinal — nausea.[14]

Nineteen ES patients who were medically evaluated in England described these reactions to EMFs: unconsciousness, fatigue, headaches, colitis (colon inflammation), neck pain, hyperactivity, dizziness, nausea, diarrhea, and extreme weakness.[4]

ES symptoms may turn off and on with the EMF exposure, there may be a delayed reaction, or a prolonged effect lasting hours or days after the exposure ceases.

The Other ES Symptom

While doing a radio show, the author was asked by a sincere talk show host whether it was possible that a person could affect electrical appliances, instead of the EMFs affecting the patient. The host explained that she knew someone who insisted that they could jumble computer disk information and distort computer screen displays just by being near them.

In England, one medical question posed to ES patients is whether appliances work all right when used by the patient.[9, 10] Apparently, some ES patients and otherwise normal folks can emit EMF signals that interfere with the electricity of appliances. A person trying to use a washing machine may find the equipment shuts off immediately after it turns on when they try to use it, while working for other operators. Television sets and remote controls are also susceptible to human EMF interference.[10]

Power line radiation is known to affect television and radio reception, and distort computer screen displays. In particular, magnetic fields from electrical power lines are known to wipe out computer disk information and affect the computer's screen display.[29] EMFs emitted by some people may do the same thing. (See also Metal Sensitivity, page 42.)

A World Health Organization publication reported that the human body emits EMFs up to 300 gigahertz (300,000,000,000 Hz), but the strength is very low (about .003 watts per square meter).[136]

The Psychology of ES

When a person first suspects that EMFs are producing their illness symptoms, voicing this concern often leads to predictable results. The new ES patient may be harshly condemned by medical doctors, scientists, their electrical utility company, their employer, friends and relatives for considering such a "crazy" idea as electricity hurting them. Their job, friendships, and marriage all become severely stressed, both by the limitations of the ES patient and by the mental trauma of trying to cope, with little emotional support.

When a primary caregiver is incapacitated, the whole family feels threatened because part of their support system is missing. Incapacity for the ES patient may mean inability to drive, cook, shop, remain in the workplace, or even use the phone due to symptoms experienced by the EMF exposures involved. If an ES person cannot stand the EMFs of a computer or television set operating in the house, the spouse and children may not well tolerate changes they need to make in their computer or television usage.

The ES patient may reduce their demands on others in order to maintain family harmony, but at the cost of their own diminished health.

Because traditional medical doctors generally know little or nothing about environmental illness, those patients with ES may hear words like psychosomatic, hypochondriac, or psychotic to "explain" their symptoms. The suggestion of seeing a psychologist or psychiatrist may be presented as a way to work through their mentally-based illness. Electrical sensitivity *is not* a mentally-based illness. (See The Causes of Electrical Sensitivity, page 20.)

If chemical sensitivity is present with ES, it is possible that brain function has been impaired due to toxic chemical exposure (neurotoxins). For example, organophosphate pesticides, known neurotoxins, reduce enzyme function and produce nervous system damage — MCS patients most severely affected are among those so exposed.[94] Petroleum-based products (plastics, solvents, pesticides, paints, etc.) are chemicals having neurological symptom reactions for many chemically sensitive patients.[86, 88] Toxic chemical exposure can damage the brain and lead to subsequent neurological symptoms upon future exposures to chemicals or EMFs. Radiation exposure from EMFs may also lead to nerve and/or brain damage, if the exposure is severe enough. So, in some cases, brain and/or nerve function may be impaired initially due to a toxic environmental source. Then, brain and/or nerve function can be further impaired by subsequent triggering exposures to offending chemicals or EMFs. Above average EMF exposure has been shown to be related to increased depression and higher suicide rates.[135]

Beyond whatever nervous system damage that may be present, several typical emotional states occur when ES patients assess their predicament:

- Anger — at doubtful family members or the employer, if work-related cause.

- Depression — can be induced by EMF exposure or by general exasperation in their situation.

- Frustration — with their condition.

- Hopelessness — when no one understands or they cannot find medical help.

- Restlessness — regarding their EMF-imposed limitations.

- Searching — that unending quest for answers, products, or therapy to help them feel better and return to a normal life.

The stress of environmental illnesses is compounded by the emotional stresses of often inadequate/inappropriate medical care, the severe social and functional limitations often imposed, and lack of understanding and emotional support from family and society in general. Family assistance and acceptance are very necessary for ES patients as well as for other environmental illness patients due to the often traumatic and life-changing events involved.

The Causes of Electrical Sensitivity

Current available information does not indicate that microbes — bacteria, virus, fungus, etc., play any initiating role in ES. Instead it is believed to be an environmental illness, caused by environmental toxins — primarily electromagnetic fields.

Important answers that need to be defined include those to the following questions:

- What types of people are getting electrical sensitivity (ES)?

- What was the initial incitant?

- Were there any pre-existing health conditions present that may have predisposed the person to become ES?

So far, available information points to three basic types of people most often getting ES:

❶ multiple chemical sensitivity (MCS) patients, ❷ computer users, ❸ people exposed to high levels of EMFs for prolonged periods of time.

Group ❶ Multiple Chemical Sensitivity Patients

MCS is a severe and debilitating form of chemical sensitivity. After becoming ill from chemical exposure, some MCS patients begin to experience further illness from EMF exposures.[8] An English medical article in Clinical Ecology notes that there seems to be a link between ES onset and exposure to certain chemicals — pesticides, in particular, are mentioned.[10] Chemical sensitivity often precedes or accompanies ES.[23, 102, 109] A general trend, based on anecdotal accounts, seems to be that the more chemically sensitive a person is, the more apt they are to be or become ES too.

Sweden's FEB group (The Association for the Electrically and VDT Injured) in Stockholm, relates that they found ES can sometimes lead to chemical sensitivity. So, one environmental illness may predispose the other.

Why are some people with chemical sensitivity having electrical sensitivity problems or the other way around?

Chemical overexposure which causes chemical sensitivity can create nervous system overload/damage from neurotoxic chemicals.[86, 106] Therefore, subsequent exposures to similar chemicals cannot be adequately tolerated. Also, fat-soluble chemicals can store in body fat and organs, remaining in the body to chronically sensitize the chemically sensitive patient to those chemicals already overexposed. So, for the chemically sensitive patient, two factors may be at work to lead to ES: nervous system (neurological) damage and toxic chemicals stored in the body.

To understand how anyone could become sensitive to electrical fields, first it is important to know that the human body is electromagnetic.[9, 33] Modern hospital equipment monitors the brain's electricity by electroencephalogram (EEG) and the heart's electricity by electrocardiogram (EKG). The brain, heart, and nervous system of the body are all highly electrical. In addition, all living cells have electrical properties.

Many of the ES symptoms are manifestations of nervous system disfunction when in the presence of EMF sources. Neurological damage from chemical overexposure may, therefore, limit a chemically sensitive person's ability to deal with EMFs due to the electrical nature of the nervous system. Induced electric currents in the body from outside EMF sources have been shown to affect the body's nervous system.[102] According to a World Health Organization publication, EMF frequencies below 100 kilohertz (100,000 Hz) are most interactive with the nervous system, while higher frequencies are more apt to produce heat effects.[136] <u>EMFs may be neurotoxic</u>. Essentially, low-frequency, low level EMFs are able to interfere with the electricity of life processes. For the ES patient, their symptoms are the clear proof of this interference.

Secondly, chemicals are each electromagnetically unique. Organic chemistry experiments use spectroscopy to identify chemicals by their absorption properties in the x-ray, microwave, infrared, ultraviolet, and visible light regions of the electromagnetic spectrum. These experiments are also called spectrum analysis.[55, 125]

Some drugs make people more sensitive to sunlight because the drugs have a high absorption rate in the visible light/ultraviolet part of the electromagnetic spectrum. The same could hold for other chemicals that have high absorption rates in other (lower frequency) parts of the spectrum. (See What Are Electromagnetic Fields (EMFs)?, page 61.)

The author's theory is that chemicals which have a high absorption rate in the lower end of the electromagnetic spectrum may induce a sensitivity to electromagnetic exposures that represent the same frequencies the specific sensitizing chemicals respond to. Toxic chemicals stored in the body could intensify the ES reaction because the stored chemicals are strongly reacting to the EMFs, and further sensitizing the patient to that EMF frequency. This could explain the frequency-specific nature of ES, from EMF interaction with stored toxic chemicals to EMF interaction with body chemistry.

A final possibility as to why chemical sensitivity can lead to ES is found in an experiment where a rabbit was injected with poison.[61] The initial result was a monitored decrease in the rabbit's liver and muscle electricity. Higher electromagnetic energy sources are known to affect lower electromagnetic energy sources. If a person's body electricity was reduced due to chemical toxin exposure, their sensitivity to external electrical fields could be increased due to their own lowered electrical field.

A vital question that has not been scientifically answered yet is "What specific chemical exposures most lead to ES?". Currently, the most neurotoxic chemicals are the more likely ES initiators, when ES has a chemical cause.

Why are some electrically sensitive becoming chemically sensitive? Possibly sensitizing oneself to EMFs that represent certain chemicals could make contact with those chemicals produce symptoms as well.

Also, nerve damage from EMFs may predispose some chemical sensitivities. Another idea is that chemical exposure was an unknown co-factor in the initial EMF sensitizing process, such as working with a computer — which has flame retardant chemicals, phenol, plastics, etc.

Group ❷ - Computer Users

Computer users as a group are an unhealthy people. Two of their major health complaints are eye problems and carpal tunnel syndrome. According to the American Optometric Association (AOA), eye strain, burning eyes, and blurred vision are among the most common eye problems reported by computer users. The AOA also cites frequent reports of headaches and neck aches among computer users.[27] Interestingly, burning eyes, headaches, and muscle/joint pain are also ES symptoms related to EMF exposure.

Computer monitors have been measured for EMF emissions and the following types reported: x-rays, ultraviolet, microwaves, visible light, infrared, radio waves, and power line radiation.[2, 40, 53, 68, 69] (See Computer Radiation Standards, page 77.)

Is carpal tunnel syndrome related to computer radiation exposure? Carpal tunnel syndrome (CTS) has a history of occurrence in occupations which require repetitive motion, such as assembly line work and chair caning. Typing at a computer requires repetitive motion too, so certainly CTS could be expected to some extent. However, CTS is now an epidemic, particularly among computer users.[93, 118] One reason may involve the use of electronic monitoring at many large employers. This type of electronic surveillance counts key strokes, the time used for lunch and breaks, and when the person arrives and leaves. Supervisors can evaluate typing speed and accuracy by computer, while the clerical employee is essentially chained to their computer monitor.

The 1987 book <u>VDT Syndrome - The Physical and Mental Trauma of Computer Work</u> written by the National Association of Working Women, Cleveland, Ohio examined 34 cases of computer-related illness. Most had carpal tunnel syndrome, 4 cases were skin rash, 5 ganglion cysts, and a 34-year-old with cataracts. They also noted that significant numbers of workers were reporting nausea, dizziness, or constant exhaustion related to their computer work.[93]

One clue that computer radiation may be a co-factor in the CTS epidemic is found in a book regarding EMFs published by the World Health Organization, Geneva, Switzerland.[136] Relating to frequencies in the 300 Hertz to 300 gigahertz (300 - 300,000,000,000 Hz) range, it states that energy (EMF) absorption increases at the neck, legs, and wrists. The reason absorption increases at these points is because they are areas of smaller cross sections in the body. Absorption rates in the wrists and ankles can exceed the overall body EMF absorption rate by as much as 300 times depending on the EMF frequency. No specific frequencies were given relating to these high energy depositions. However, the two primary frequency ranges monitored for computer monitor EMF standards are at 5 Hertz - 2 kilohertz (5 - 2,000 Hz) and 2 kilohertz - 400 kilohertz (2,000 - 400,000 Hz).[121] Therefore, both of these ranges are within the World Health Organization's frequencies under discussion above.

The two primary EMF frequencies of computer monitors are at 60 Hertz (U.S.) and at about 20 kilohertz (20,000 Hz).[123] However, harmonics spread the range of frequency emissions. Computer monitors also have higher frequency emissions due to the cathode ray tube technology.

Other health complaints suspected of being related to the computer user's radiation exposure are epilepsy, cancer, miscarriages, birth defects, cataracts, and electrical sensitivity.[2, 40, 53, 68, 69] For example, in 1992 the American Journal of Epidemiology reported the results of a study from Finland regarding miscarriages among female computer users. Clerical workers at three Finnish companies were the study subjects. One hundred ninety one miscarriage cases were compared with 394 births. Those who used a computer monitor emitting a high level of extremely low frequency (ELF) magnetic fields (more than 9 milligauss at 20 inches from the screen) were 3.4 times as likely to have had a miscarriage instead of a live birth compared with computer users having ELF magnetic field exposure levels below 4 milligauss.[81] Miscarriage clusters among office workers have been documented worldwide.[2]

In 1992, Sweden's National Institute of Occupational Health surveyed five major employers in Sweden and found one of every seven employees (out of 731 total) experiencing symptoms of ES.[23] Skin complaints were the most common, with ear/nose/throat symptoms second, followed by eye problems, mouth problems, and nervous complaints. Recorded nervous symptoms were fatigue, weakness, headaches, dizziness, prickly sensations, depression, heart rate fluctuations, forgetfulness, perspiration, and breathing problems. In one group of workers, six out of ten were allergic to chemicals in addition to ES symptoms. Some workers were also light sensitive, so sensitive that they could not be near anything brighter than candlelight.

Skin problems for these studied workers were two times as common if workers used computers more than four hours daily. Also, the study noted that workers using higher magnetic field emitting computer monitors had more skin problems than computer users with lower magnetic field monitors.[23]

A 1994 medical evaluation of ES skin problems done by the Experimental Dermatology Unit, Department of Neuroscience, Karolinska Institute, Stockholm,

Sweden concluded "...that certain paramount and profound changes in the dermis and epidermis take place..." upon exposure to television set EMFs.[17] Basically, in ES skin patients, immunoreactive skin cells were much greater in number than in non-ES controls. The scientist who headed this study is currently at risk of having funding discontinued. More studies like this are imperative.

In 1990, 1650 patients of Sweden's health care centers with ES-type skin symptoms were evaluated. Their symptoms were then categorized by degree of ES severity:

	Patients	Percent	Symptoms
Category 1	1150	69.7%	Skin pain at computer only.
Category 2	350	21.2%	Residual skin pain after computer use or other electrical exposure.
Category 3	90	5.5%	General electrical sensitivity.

— Sick leave or work transfer was required. —

	Patients	Percent	Symptoms
Category 4	60	3.6%	Serious ES symptoms, even at home.

—Severe work and lifestyle limitations. —

Source: Work with Display Units 92[12]

Preliminary results of a recent survey of The Swedish Union of Clerical and Technical Employees in Industry (SIF), Stockholm, Sweden found the following indications of possible ES among 1,694 members who responded:

• For men, the most frequent symptom is eye problems. A large percentage of respondents were young men under age 36.

- For women, more than one out of three have eye symptoms, joint pain, and skin problems.[16]

Sweden's Association for the Electrically and VDT Injured survey results indicate that 90 % of their members report computer usage as their cause of ES.[14]

Pre-survey indications in the United States show ES men predominantly former computer users, not necessarily MCS. United States MCS patients appear to be mostly women, so United States patients with MCS who get ES are often women, but not exclusively.

How could computers cause electrical sensitivity? Prolonged exposure to the EMF emissions of computer monitors is one strong possibility, with chemical emissions, particularly with brand-new equipment, a possible co-factor.[23, 69]

According to Sweden's Karolinska Hospital letter published in England's medical journal Lancet, computer skin rashes were suspected of being linked to static electricity drawing airborne particles to the computer user's face. When static electricity shielding was applied to the screen, some skin rash patients reported improvement while others did not improve.[3]

Beyond static electricity, the alternating magnetic fields create electric currents in electrical conductors, like the body. These electric currents create heat (in the body) which is called *induction heating* in electronics.[62] In other words, currents create heat, as does energy absorption at higher EMF frequencies. Currents tend to flow more on the surface of the conductor as the EMF frequency increases. In electronics, when the current flows more on the surface of a conductor, it is called a *skin effect*.[62, 89] Computer users are getting a *skin effect* also, it appears. What EMF frequencies and/or chemical offgassing factors produce their skin problems has not been scientifically determined. However, it seems clear that heating the skin with induced electric currents could lead to prolonged skin inflammation.

The scientific theory that most induced currents currently are non-thermal (meaning you cannot feel the heat) is no guarantee that they are safe. Excessive heating of the body by high levels of radio frequency EMFs, above computer monitor levels, has been shown to produce cataracts, birth defects, and miscarriage, according to World Health Organization and International Radiation Protection Association publications.[71, 136] The eye is especially sensitive to heat because it does not have the blood supply to dissipate heat.[136] This fact is one reason why eye problems are being discussed in relation to computer radiation exposure. Induced heating of the computer user's wrists could possibly lead to tendon inflammation that compresses the median nerve, producing carpal tunnel syndrome. Office ergonomics are helpful to ensure repetitive motion is minimized and posture is comfortable, but office assessments need to consider both ergonomic and environmental factors.[122, 123]

Group ❸ - People Exposed to High Levels of EMFs for Prolonged Periods of Time

A third way ES can occur is by being exposed to higher levels of, or different types of, EMF than you had previously been exposed to.[34] This new EMF exposure can produce symptoms when your body does not adjust to the EMF stress.

Fluorescent lights and energy-efficient bulbs are common culprits for ES reactions.[14, 23, 92, 109] Sometimes they appear to be the initial incitant that begins the ES process.[109]

Denmark's electrically and VDT injured group, EBD, says that most of their members were computer users or in an electrical environment which caused the ES. Their group has about forty-four members. Denmark's group is arranging the 2nd Copenhagen Conference on Electromagnetic Hypersensitivity, an international scientific conference, to be held at the University of Copenhagen, Denmark on May 22 -23, 1995.[1]

In Sweden during the late 1980's, about fifty young technicians at one employer, Ellemtel, became electrically sensitive. They were computer users and had additional potential EMF exposure from three microwave transmitters situated on the employer's roof.[19, 20] ES symptoms one worker experiences are extreme weakness and a skin burning sensation like a severe sunburn when EMF exposed. This worker now, of necessity, has an EMF shielded room at work that he shares with other ES workers. At home he occupies one shielded room, shielded with 1,100 pounds of iron sheeting with two layers of silicone.

His symptoms began shortly after using high-resolution color computer monitors from England. He then began experiencing burning and stinging on his face at the computer. The symptom reactions spread to other EMF exposures: fluorescent lights, cars, and electrical appliances in general. He now has "hot spots" and red patches all over his body, which intensify near anything electrical. He was also light sensitive early on, but that symptom left over time.[19]

A survey of United States ES patients is being prepared, in part to better define the causes of ES. Currently, any EMF source could be considered a potential ES initiator.

In the future, all electrical appliances must be required to be shielded to reduce EMFs, perhaps by amending The Radiation Control for Health and Safety Act of 1968 designed originally to legislate reduction of x-ray emissions from television sets.

In reality, everyone is electrically sensitive at the cellular level; with some people the EMF reactions are less subtle than with others. The EMF cancer issue is a prime example.

Cancer, in its present form, could even be interpreted as an environmental illness, a cause external, not from a contagious source, but of an environmental basis. Cancer itself could also be loosely interpreted as ES, as EMFs in some studies promote the growth of cancer cells — cancer sensitive to and promoted by the electrical fields.[128] So, we could say that even though cancer patients may not feel EMFs, cancer is electrically sensitive. Further, you could surmise that everyone is essentially electrically sensitive, whether they notice it or not. The body's cells, nerves and organs are sensitive to external electromagnetic fields. When an ES patient tells you that the refrigerator's EMFs are biologically active at 20 feet, the implications of the biological activity of EMFs is in question for everyone.

What electromagnetic field sources are most difficult for ES
patients to contend with?

Sensitivity varies depending upon the degree of illness, mild or severe, the uniqueness of the individual, and the variety of products and sources available.

Generally, computer monitors, television sets, motorized appliances, fluorescent lights, transformers, power lines and household wiring problems pose the greatest difficulties for the ES patient. Anything electrical is a potential health threat to those with ES.

Is electrical sensitivity more severe at some specific electromagnetic frequency
exposures as opposed to others?

No known large-scale studies have been attempted yet to determine what frequencies ES patients may be most sensitive to. However, in a study of 100 ES patients, sixteen clearly symptom-responsive patients had the following frequency reactions:

1	Hertz	75%
2.5	Hertz	75%
5	Hertz	69%
10	Hertz	69%
20	Hertz	69%
10	kilohertz (10,000 Hz)	69%

Other frequencies tested (.1 Hertz (Hz), .5 Hz, 40 Hz, 50 Hz, 60 Hz, 100 Hz, 1 kilohertz (kHz) (1,000 Hz), 5 kHz, 20 kHz, 35 kHz, 50 kHz, 75 kHz, 100 kHz, 1 megahertz (MHz) (1,000,000 Hz), and 5 MHz) recorded less than a 69% response rate. Of these frequencies, the lowest responses (31%) occurred at .1 Hz, 35 kHz and 5 MHz. None of the 16 patients recorded ES reactions to all of the frequencies; the response range was between 1 and 19 positive responses out of the 21 frequencies tested.[102]

Of interest are the brain wave frequencies:

Brain Wave	Mental State	Frequency Range
Delta	Asleep	.5 Hz - 4 Hz
Theta	Deeply Relaxed	4 Hz - 7 Hz
Alpha	Relaxed	8 Hz - 12 Hz
Beta	Awake	14 Hz - 30 Hz

Source: Encyclopedia of Science and Technology[88]

Therefore, frequencies within the range of .5 Hz - 30 Hz would be expected to show the most neurological responses because they are simulating brain frequencies. The study above shows that five of the six highest response rates were from frequencies within the range of brain frequencies. And, indeed, neurological reactions were the ES symptoms reported most in this study — sleepiness, headache, dizziness, tingling, unconsciousness.[102]

Power line frequency is 60 Hz in the United States and 50 Hz in Europe. Computer monitors have two primary frequencies: power line frequency and very low frequency (VLF) of about 20 kHz (20,000 Hz). Fifty-six percent of the above study's patients had ES responses at 20 kHz, 63% responded to 60 Hz, and 50% responded to 50 Hz.[102]

Case Examples of Types of ES

The examples below are real people facing enormous health, legal, medical and social problems due to their electrical sensitivity.

Some are multiple chemical sensitivity (MCS) patients who developed ES, others are ES only. Computers and other EMF factors account for the non-MCS cases. The author has talked to or corresponded with all of the following ES patients:

Case 1: Pension administrator in office setting develops a heat reaction on her legs. Her office is next door to the computer room, where four computer monitors run all day. The heat reaction occurs at her desk, diminishes in other offices and disappears at home. Administrator's office is moved due to management's reallocation of office space. Heat on legs disappears and her assistant does all of the computer room work.

New office policy indicates that all employees will be getting a computer in their office. The office across the hall gets a new computer monitor. Heat reaction begins again and intensifies to extreme level. Administrator discusses the problem with her supervisor and supervisor agrees to move employee to a new office, away from computers. Employee gives three weeks notice to quit her job.

New office is away from both the computer room and the old office. However, unknown to the administrator, a new employee moved a color monitor against the wall behind her new office. The heat reaction begins again in the new office — a search of the area finds new employee's computer. Computer cannot be moved; administrator moves to the conference room to work. Her dermatologist, when asked whether computer radiation could be a factor contributing to the rash and heat reactions, answered, "Ridiculous".

Conference room was not always accessible and her office was unacceptable so employee requested to do work from home until her date of termination came. Supervisor agreed. Employee quit and has since had to quit two other jobs due to EMF exposure problems also. Worker's compensation was declined. Employee is not 100% disabled so cannot qualify for Social Security. Her business career is over, as she cannot be within two rooms of a computer monitor without developing "hot spots" on her legs. Her only hope is finding a non-computer job in an EMF safe location, but nearly *everyone* has a computer. She now also reacts to television sets, most motorized appliances, and other EMF sources.

Type of ES patient: Computer user

Case 2: Employee is an insurance agent who begins to feel ill at work. A doctor determines that the fluorescent lights are bothering her. She can work from home, but cannot comfortably work at the office anymore. When she leaves home, she must wear dark sunglasses indoors, even on cloudy days, to protect her eyes from fluorescent lighting in malls, stores, and most commercial and governmental buildings everywhere. When required to be in a fluorescent lighting environment, she needs to be close to the windows to get as much of the natural light as possible, while wearing her sunglasses indoors. Years after the author met this person, she wonders whether fluorescent full-spectrum lighting at the office would allow the insurance agent to return to the office? The question is what frequencies were bothering her — and would full-spectrum lighting benefit her or are EMF frequencies that fluorescent lighting has in general making her ill, regardless of the bulb.

Type of ES patient: Incitant was EMF exposure to fluorescent lights

Case 3: Person with a Master's degree developed multiple chemical sensitivity followed by electrical sensitivity, in 1981. Her MCS has diminished somewhat over time, but she develops movement disorder, then paralysis from some EMF exposures. She says it mimics some of cerebral palsy or Parkinson's disease symptoms. In addition, severe seizures occur upon exposure to generators, transformers, elevators and other higher EMF emitters. One attack has been documented in an EMF video.[30] During filming, she walked up to a building — her neurologist's office — walked in, and when she got near the elevator, had a seizure. It took assistance from the video's producer to prevent her from bumping her head during the seizure and to get her out of the building, as she could not function then without help. She maintains that EMFs are biologically active and that we, as a society, must reduce EMF exposure to protect everyone's health.

She has Social Security benefits due to her functional disabilities near EMF, chemical, and smoke exposures.

Type of ES patient: Person with MCS

Case 4: A woman with chronic fatigue syndrome and multiple chemical sensitivity wrote the author about a reaction she had with her new computer monitor. Right after she started using the computer, she developed a skin rash on her face and her eyelids were puffy. She used the words "burning sensations" to describe her symptoms.

After she discontinued using the computer and returned it to the dealer, her rash and puffy eyelid symptoms went away.

Type of ES patient: MCS patient and computer user

Case 5: A teacher wrote to the author about having multiple chemical sensitivity, from which both ES and sun sensitivity also developed. She is housebound, as she cannot get into any car due to the reactions which develop there. A car can have many chemical exposures, plus EMF exposure due to the motor running. Also, ES patients can become sensitive to metals which make them react to metal enclosures, like cars. (See Metal Sensitivity, page 42.) When an ES patient is reacting to something like chemical or EMF exposure, they can produce electrical signals that bounce back to them when in metal enclosures.[9, 10] These rebounding signals make the ES patient's reactions worse. Essentially, then we have a type of autoimmune problem happening in these cases; the patient is allergic to themselves, or their own signals. Also, this patient mentioned not being able to touch her refrigerator, stove, or even the water faucet without getting a reaction.

In England, a published report on ES patient testing noted that it is not uncommon for ES patients to say they feel electric shocks when touching home or office appliances.[4]

This patient cannot use a phone, due to the EMF exposure.

Type of ES patient: MCS patient

Case 6: A male ES patient wrote the author, noting an intense reaction in his heart when applying the hand-brake on his bicycle. He may be reacting to the sound frequency of the brake upon application. (See Sound Sensitivity, page 37.) Rather than sound sensitivity, he felt it was a type of metal sensitivity and that his aluminum brake was attracting and transmitting EMFs. One or both factors may be what makes him react.

Type of ES patient: Computer user

Case 7: Computer user describes a skin rash on her legs. Works in close quarters with many other computer users. Wonders whether the computer disk drive box on the floor at her feet is contributing to her skin rash. The dermatologist does not know what is causing her rash and cannot help her.

This employee notices that she feels heat sensations on her chest when using the office's microwave oven. She is concerned about the oven and has expressed her concern to co-workers, who do not notice anything wrong. She told the author that several co-workers are pregnant and use that microwave oven.

Employers take note: Have you checked the microwave emissions of the office microwave oven lately? Older ovens can lose their door seal or get dirt and crumbs in the seal, causing microwaves to escape. Microwaves cook food. This is intense radiation that can permanently harm an unsuspecting person. Microwave ovens should never be operated with the door open. Microwave ovens are not suitable for children to use. In addition to possible microwave emissions, many microwave ovens emit significant power-line type radiation at operating distances. (See Magnetic Field Readings from Common Home Appliances, page 82.)

Microwave ovens should be periodically checked for microwave emissions and repaired/cleaned/replaced, if emitting unacceptable microwave levels based upon the measurements. ES patients would do best to avoid microwave ovens altogether, due to the EMF exposures.

Type of ES patient: Computer user and high EMF exposure to microwave oven.

Case 8: A nurse administrator in an out-patient clinic, also a computer user, began having nausea, headaches, intense fatigue, difficulty concentrating, chest pain, and dizziness at work the week energy efficient lighting was installed. Also that week, her building was evacuated due to a chemical smell.

She developed such severe reactions that she went on sick leave and camped out in the pasture near her home to avoid EMF and chemical exposures as much as possible.

Food sensitivities, allergic-type reactions to certain foods, also became apparent. Sugar, alcohol, and caffeine are particularly troublesome for her. Wearing metal jewelry, watches, and being in close proximity to batteries accelerated the electrical sensitivity.

This nurse has been frustrated that conventional medical doctors seen early in her search for answers did not understand her ES and their formal remarks of disbelief cost her health insurance denials and worker's compensation benefit denials.

Months after the onset of symptoms and these denials for benefits, she did find and receive care from an M.D. specializing in Environmental Medicine. This physician is familiar with electrical and chemical sensitivities and treated her for chemical sensitivities.

When her sick leave and vacation time were used up, she had to return to work or risk forfeiting her job altogether, with no other means of income support. Her environmental medicine physician allowed a trial return to work; she relocated to another area of the building.

While managing her symptoms through avoidance of EMFs and chemicals, and treatment for MCS, she can function to some extent. Symptoms flare-up, however, when leaving her new office to communicate with other employees in her normal work capacity.

Type of ES patient: Computer user, high EMF exposure to new fluorescent lighting, and possible MCS co-factor.

There is a great need for acceptance of ES on federal and state levels to insure that ES patients can get much needed services now denied them: Social Security, worker's compensation, protection under the Americans with Disabilities Act, low-income housing needs, removal of electronic barriers for access to public buildings, etc.

Opinion on Symptom Effects

Many complaints of ES surround heat reactions, feelings of pain and pressure, and organ disfunction. Some published ES reactions were severe enough to cause coma or death.[4]

The author believes that body weaknesses, in whatever form, are the areas/organs most susceptible to ES reactions in an ES patient. When EMF exposure produces induced electric currents, where these currents are impeded may be where the heat reactions, pain, pressure, and organ disfunction are occurring. Chemical or metal concentrations in those areas may be a factor. Cancer, pre-cancerous conditions, chemical damage, burns (from fire, sun, or radiation), injuries, metal implants, etc., could be other pre-existing conditions leading to localized ES reactions at that site. "Hot spots" mentioned in connection with ES may have been initiated by radiation exposure concentrated near or projected from electrical equipment; *hot spot* is an electronics term.[62] Cancer is medically known to limit dissipation of heat, which may lead to some ES-type heat effects from EMF exposure.[63] The nervous system of the body may in some cases become irreparably damaged by chemicals or radiation exposure, resulting in severe neurological ES symptoms. This theory needs to be scientifically explored.

Diagnosis of Electrical Sensitivity

A difficult part of any environmental illness is sorting out the possible causes of reactions. Allergic-type reactions to foods, chemicals, and electromagnetic fields all need to be isolated. Further sensitivities may involve symptom responses to sounds, changes in the weather, sunlight, all light, metals, watches, other people, amalgam dental fillings, and/or geopathic stress.

<u>Sound Sensitivity</u>:

Sound sensitivity means that a person is extremely sensitive to sound. Certain pitches or frequencies of audible or inaudible sound cause the person feelings of stress or ill health. With sound sensitivity, noises may sound louder and more piercing than they really are.

Sound waves are a type of radiation called acoustic radiation. In medicine, ultrasound can take pictures of soft tissues, instead of x-rays. Ultrasound means sound above the level we can hear.

Some types of electronic products, like computer monitors, produce ultrasound.[2, 53] A number of people have reported hearing a sound at computer monitors that others cannot hear. They may be detecting sound frequencies that people normally cannot hear. Sweden's TCO labor union notes that high frequency sound at and above 16 kHz (16,000 Hz) can be a troublesome noise, particularly for young computer users.[123]

Of interest is the fact that carpal tunnel syndrome (CTS) is medically known to be at higher risk where repetitive motion, temperature, force, or vibration are extreme. Ultrasound is a type of mechanical vibration occurring at the computer monitor.[71] All radiation, including ultrasound, produces heat. The most common way CTS occurs is by tendon inflammation which compresses the median nerve. Heating the body by ultrasound or other radiation may possibly lead to such inflammation or contribute to it.

Headaches and nausea have been reported by the International Radiation Protection Association (IRPA) when certain people use ultrasound cleaning tanks or approach commercial buildings having an ultrasound intrusion alarm. Fatigue, tinnitus, and ear pressure are other symptoms the IRPA also related to ultrasound exposures. Consumer products that use ultrasound include garage door openers, some alarms, some remote controls, electronic pest repellent, and guidance devices for the blind.[71]

You can become sensitive to sounds that you cannot hear because sound waves, like EMFs, are a stressor to the human body. When a person becomes ES it is possible that they may become sound sensitive also. Avoidance is the key with all sensitivities. If you suspect that something is bothering you, avoid the possible incitant to see if that avoidance helps you feel better.

Weather Sensitivity:

If you can feel a storm coming, you may be weather sensitive. Certain people feel ill or stressful a day or two before a weather front arrives, then feel relaxed and relieved when the storm comes. With thunderstorms, the stressful feeling may not pass until the weather calms.[72, 115]

Weather changes affect the concentration of positive and negative ions in the air. Ions are microscopic charged particles. Positive ions are considered stressful; negative ions are relaxing. Air normally has a positive and negative ion mix, with mountains, forests, ocean beaches and waterfalls having a high level of negative ions in the surrounding air. Negative ions are related to friction between the air's moisture content and the wind, while positive ions are related to dry, warm winds.[51]

In the 1970's, Felix Sulman, M.D. at Hebrew University, Hadassah Medical Center in Jerusalem, Israel found that the neurotransmitter serotonin is released in the body after high positive ion exposure. About the same time, Albert Krueger, M.D. at the University of California, Berkeley, California also found that positive ions were unhealthy in large concentrations while negative ions did not cause the release of serotonin in the body.[51, 72, 115]

Dr. Sulman's work demonstrated that serotonin release caused several unhealthy symptoms: headache, problems swallowing, dizziness, itchy eyes, breathing problems, and itchy/burning/dry nose.[51]

Before a storm, positive ion concentrations increase while after the storm negative ions increase. Weather sensitive people may be reacting to the higher concentration of positive ions.[105] Another possibility that ES patients may notice is ill health just prior to and during a thunder and lightning storm. The changes in air electricity other than ions can be a factor in causing ES-type reactions to weather. Areas with frequent or harsh storms can be quite disturbing to the weather sensitive person. A mild climate with little seasonal change may help to minimize the weather "spells".

Areas in the United States that have particularly harsh positive-ion creating winds are the Chinook winds in Montana and the Santa Ana winds in California. Foreign winds considered unhealthy to the weather sensitive include the Foehn in Germany and Switzerland and the Sharav in Israel, among others.[51, 72, 115]

Indoor environments are also often sources of high positive ion counts due to electrical equipment, poor air quality, and inadequate ventilation. Office environments are prime examples of places where serotonin release symptoms can occur.[51, 115]

Sun Sensitivity:

Sun sensitivity could present itself as a general feeling of illness, or an intense heat reaction on the skin, when sun exposed. Sun sensitivity differs from light sensitivity. With light sensitivity a person is ill when reacting to *any* form of light, not just sunlight.

Some pharmaceutical drugs can cause a temporary sun sensitivity. The prescribing doctor would generally advise you to stay out of the sun while taking that type of drug to prevent a severe sun reaction. Chemically induced sun sensitivity may diminish or disappear over time.

Sun sensitivity is really an EMF sensitivity because visible light and ultraviolet are part of the electromagnetic spectrum.

Anecdotal accounts of sun sensitivity indicate a relationship to silicone breast implants that leaked. This could be true as silicone is related to silicon which is used in energy solar collector panels because of its sun sensitivity.

Severe sunburns may lead to sun sensitivity of the skin where the burn occurred. Sunburn induced sun sensitivity can also lead to localized ES at the damaged skin sites. Induction heating by EMFs can cause significant discomfort where heat dissipation is restricted, like old sunburns.

Another way sun sensitivity appears is with the autoimmune disease called *lupus*. Symptoms of lupus include any or several of the following: achy joints, fever over 100 degrees Fahrenheit, arthritis, prolonged fatigue, skin rashes, anemia, kidney problems, chest pain when deep breathing, butterfly-shaped rash across cheeks and nose, sun sensitivity, hair loss, white or blue color in cold fingers, seizures and/or mouth ulcers.[83] An important cross-over are three of these symptoms which are not unusual in ES patients who were prior computer users: skin rashes, butterfly-shaped rash across the cheeks and nose, and sun sensitivity. When lupus is limited to the skin it is called discoid lupus. When other organs are also involved the term is systemic lupus erythematosus (SLE). Lupus is fairly common, but not well known. One person of every 185 in the United States has lupus.[83] MCS patients may have lupus (SLE) symptoms; lupus sometimes has chemical exposure beginnings, particularly trichloroethylene (TCE), a solvent.[82, 94] When isolating factors involved in a person's illness, the doctor may need to consider whether the diagnosis of lupus fits the symptom picture. With discoid lupus, sunscreens are used to reduce rashes when sun exposed. Generally, lupus patients are not known to be electrically sensitive, only sun sensitive. However, discoid lupus patients do react to ultraviolet exposures from fluorescent lights and older television sets.[25]

Sun sensitivity may also be related to the disease porphyria. (See Light Sensitivity below.)

Staying out of the sun is your best defense against sun sensitivity reactions. One way to do this without becoming housebound is to relocate to an area with lots of cloudy days, like Oregon or Washington. Severely sun sensitive people may not be able to comfortably go outside even on cloudy days.

Light Sensitivity:

The inability to tolerate light — daylight and artificial light — is called photosensitivity. Visible light is a form of electromagnetic radiation. So, light sensitivity could loosely be defined as a form of ES. Photosensitivity sometimes accompanies ES, particularly in the severe forms of ES.[19, 23] Light sensitivity generally involves either nerve stimulation making the eyes squint or hurt when light exposed, or skin problems manifesting when light exposed — pain, burning, etc.

Porphyria is a disease which can produce both sun and light sensitivity. Toxic chemical exposure is one suspected cause of porphyria; however, traditionally it has been considered a genetic disease of enzyme deficiencies. Some multiple chemical sensitivity symptoms may overlap porphyria symptoms. According to the Swedish National Institute of Occupational Health in Solna, porphyrinuria (porphyrin present in the urine) is due to a malfunction in hemoglobin breakdown.[23] An important point to note here may be that chemical pollutants called nitrates from fertilizers can change the chemical structure of hemoglobin, which prevents hemoglobin from refreshing the body's tissues with oxygen.[7] Perhaps nitrate exposure in particular is a factor in the development of porphyria. Neurological and skin diseases are a common result of porphyria.

Photosensitivity is known to be related to exposure to coal tar derivatives, because coal tars are active in the visible light range of the electromagnetic spectrum as determined by spectrum analysis.[114]

ES patients with the most severe light sensitivity live in near darkness, as they cannot tolerate anything brighter than a candle without ES symptoms.[23]

An opposite reaction to light, when patients live in areas with diminished sunlight, especially in winter, is called Seasonal Affective Disorder (SAD). SAD is characterized by depression on cloudy days that dissipates in sunny weather. Where sunny weather is sparse, full-spectrum light bulbs have been used therapeutically to alleviate SAD symptoms.[36, 70] Full-spectrum lighting simulates sunlight. Incandescent light bulbs, in general, have less harmful electromagnetic emissions than the fluorescent bulbs. SAD has not specifically been shown to be related to ES, as with ES, the avoidance of various energy fields is necessary.

Epileptic seizures can occur from exposure to flickering lights, strobe lights, fluorescent lights, computer screens or other pulsating lights.[2, 92] When epilepsy is light-related, it is called photosensitive epilepsy. Young children using video games have developed this epilepsy worldwide. Photosensitive epilepsy is reportedly caused by brain wave interference from flash frequencies between 10 Hz - 43 Hz.[2]

Metal Sensitivity:

Electrically sensitive patients often are sensitive to metals. Metals have absorptive and reflective properties in the presence of EMFs. Metal can act like an antenna to attract and send EMFs. This antenna-effect is particularly noteworthy with metal-containing items worn on or in the body: jewelry, glasses, dental fillings, and surgically implanted pins and plates. (See Amalgam Dental Fillings, page 44.) ES patients have had reported difficulties with all metals listed above.[2, 9, 10] An ES patient's environmental medicine doctor might feel it necessary to do allergy testing for "allergic" reactions to liquid injections representing basic metals the person may be in contact with internally or externally.

A side issue here is whether the person has toxic metal (aluminum, lead, mercury, etc.) overexposure as represented by a blood test. Metal storage in the body might be a reason for ES in some cases, but this point has not been scientifically demonstrated as yet. Working with metals, metals in drinking water, metal dental fillings, metal cookware and copper (or other metal) water pipes in the home can all contribute to the presence of metal in the patient's body.[105] About 10% of MCS patients at the Environmental Health Center, Dallas, Texas had intracellular metal deposits.[94]

Patients being allergy tested in rooms with porcelain-on-steel walls to diminish chemical sensitivity reactions may have reactions due to the metal walls. In England, it has been reported that medical tests of ES patients may require placing a glass bottle/plastic bucket of salt water in the testing room to help absorb electromagnetic fields in the room that are bouncing off the walls. (One handful of table salt to two gallons of water was used.)[10]

Everyone is electromagnetic. When ES patients are having an environmental reaction, they emit EMF signals. It is these signals that can bounce back to them and further make them sick because of the presence of the metal walls. Metal enclosures such as automobiles may also contribute to their symptoms for the same reason.[9, 10]

If you find that your house has high EMFs and you move or correct the problem, you may find that your bed's mattress springs have been magnetized by the EMF exposure. Checking the mattress with a compass should show the compass needle pointing only north, otherwise magnetic disturbances exist in the metal springs. Metal bed frames can produce magnetic and EMF attraction problems too. A way to avoid metals when sleeping is by using a natural futon mattress with a wooden bed frame.[39]

Reactions to Watches:

People with ES can become so sensitive that they cannot wear a watch, particularly battery-operated LCDs or metal watches. The battery-operated watches and LCD watches have an energy that can be troublesome for the ES patient.[105] The metal watch could be a problem because of the metal's EMF attraction/reflection problems.

An interesting counterpoise is when someone cannot wear watches at all because watches stop working or break when that person wears them. A United States medical survey by Melvin Morse, M.D. at the University of Washington questioned over 400 people who claimed to have had a near-death experience as a child (seeing a "light" or tunnel while unconscious in an extremely ill state). One question they were asked was whether they had problems with lights, electrical appliances, or watches. More than 25% of the respondents reported that they could not wear watches because the watches stop when worn.[90] Perhaps a case of strong EMFs from the person is affecting the watch. (See The Other ES Symptom, page 18.)

Reactions to People:

Under medical testing and evaluation of ES patients in England, it was reported that if several patients were tested for EMF reactions in the same room, it could become difficult to ascertain whether the person was reacting to the EMF frequency being tested or to other patients reacting in the room. Patients who are reacting to environmental chemicals or EMFs can send out strong EMF signals during the reaction phase. Also, some people can emit signals that disturb an ES patient without being ill themselves.[9, 10]

Reactions to Amalgam Dental Fillings:

Silver amalgam fillings are made with silver and mercury. Mercury is a well-known poison, but silver amalgam fillings have not been conclusively scientifically accepted as poisonous. However, silver amalgam fillings are sometimes related to MCS symptoms.[106] ES patients who have metal dental fillings may experience tooth or mouth pain due to EMF exposure. Also, the metal mixed with EMFs cause mercury to be released, particularly if other metal, like gold, is also in the mouth. The combination of metals produces an electrochemical reaction, releasing the mercury from the amalgam fillings.[13]

A booklet from Sweden's FEB group documents a study performed by an engineer and an orthodontist that revealed computer monitor radiation exposure can increase the release of mercury in silver amalgam dental fillings.[14]

Some people sick from their amalgam fillings have them removed and replaced with non-metallic fillings. ES patients from Sweden have reported that replacing amalgam fillings with non-metal ones makes them feel better in some cases.[1]

Having the amalgam fillings removed may make the patient sicker at first due to mercury leakage when removal occurs. Removal may need to be done in stages if the person is very ill.[13]

The body can store mercury accumulations after the amalgam fillings are removed. One Swedish man recovering after having his amalgam fillings replaced reportedly takes the mineral selenium daily to help remove any residual mercury contamination in his body.[13] Selenium has been shown to help the body detoxify heavy metals like mercury.[28]

Whether amalgam fillings are a cause of ES is not clear but it is certainly not the only cause. ES has occurred in patients who do not have *any* type of dental fillings.

A medical notation from the homeopathy sector of alternative medicine indicates that homeopathic mercurius solubilis is sometimes used by homeopaths treating patients ill from their amalgam dental fillings.[103]

Reactions to Geopathic Stress:

Geopathic stress means natural magnetic field gradients in the earth — energy disturbances in the earth itself, relating to soil composition, subterranean water, magnetic field abnormalities within the earth, etc.

Some people believe that prolonged exposure to geopathic stress can lead to cancer and other illnesses.[5, 6, 38]

Geopathic stress may be troublesome for some ES patients as an article from England regarding ES diagnosis notes that geopathic stress at the testing site may affect testing results.[10]

How do you find geopathic stress? The question is not easily answered. One possible way to locate geopathic stress is by dowsing.[38] Does dowsing really work? Some natural gas utility companies in the United States have used dowsing to locate their hidden gas lines. Dowsing is not 100% reliable but it can work to locate underground springs, buried utility lines and other disturbances in the earth that create magnetic field gradients.[32] The person holding the dowsing rod(s) appears to be responding to very small changes in the earth's magnetic field scientifically measurable by magnetometers.[38, 67]

In diagnosing ES, a medical doctor who specializes in environmental medicine is often helpful to sort out the possibilities with the ES patient. (See Medical Resources, page 72.) Determination of specific food and chemical incitants may also be helpful by allergy testing. One scientific evaluation of MCS patients noted that 80% also had food sensitivities, and 90% had water sensitivity, requiring purified or spring water.[94] Diseases that are common to environmental illness patients should be watched for by the attending doctor — chronic fatigue syndrome, candida yeast, Epstein-Barr Virus, etc.[47] Fatigue is common with any environmental illness. Stress factors accompanying the illness may need attention also.

Electrical sensitivity is difficult to diagnose as the connection between the cause of the patient's symptoms and their symptom picture may not be easily apparent. Therefore, their doctor may be treating the symptoms without an understanding of the symptoms' cause.

Keeping the prior list of other/related sensitivities in mind, EMF provocation testing can reveal ES cause and effect responses to disturbing EMF frequency exposures.

In Sweden, EMF provocation testing was done in an electrically insulated room. Random EMF exposures were at 50 Hz and 20 kHz (20,000 Hz) (the two primary computer frequencies in Sweden). Fourteen ES subjects were tested by scientists from the National Institute of Occupational Health, Solna and Sweden's National Institute of Radiation. This joint study found better than random results from three of the test subjects, who individually indicated when the EMF exposure was on and when it was off.[23]

Why all of the ES subjects did not have better than random test results could be because of delayed reactions, prolonged effects, and the fact that some ES patients and symptoms have a more clearly apparent cause and effect pattern in the presence of EMFs. For example, the onset and diminishment of pain reactions — headache, tooth/mouth pain, muscle/joint pain, skin burning, pressure in the ears, etc. — may be more clearly apparent to the patient than the onset and diminishment of fatigue, sleepiness, depression, etc.

In England, at least sixty ES patients have been medically tested for EMF sensitivities by the use of a frequency oscillator with a range of .1 Hz - 10 MHz (10,000,000 Hz). EMF leakage from the oscillator is used to determine what frequencies the person is sensitive to. Their symptoms can be clinically neutralized by switching the EMF exposure to a non-reactive or less-reactive frequency. Therefore, cause and effect relationships between EMFs and ES patients can be clearly shown. These test results proved that what could be induced electromagnetically, could be neutralized electromagnetically, just as in allergy testing chemical reactions are induced and neutralized chemically. Also, they found that reactions chemically induced could be neutralized electromagnetically at a non-reactive frequency. Further, reactive EMF frequencies could sometimes be chemically neutralized by injection of the patient's individualized non-reactive chemical antigen usually used to reduce chemical or food allergies. Some ES patients reported correspondingly similar EMF reactions as with their food or chemical allergies. They noted having "a perfume reaction", "a milk reaction" or even "a thunderstorm reaction" from specific EMF-reactive frequencies.[4, 9, 10]

Because ES patients could not easily be EMF neutralized with non-reactive EMFs at home, another way of neutralizing EMF reactions was sought. An experiment was conducted by placing vials of pure water in the patient's EMF neutralizing frequency for five to fifteen minutes. Later, when the ES patient held the vial of charged water, their reactions were neutralized just as if the oscillator was turned on to that neutralizing EMF frequency. Initial findings show that the charged water will hold its charge at least two weeks, so patients can take vials home for use when an EMF reaction makes them ill.[4, 10]

One concern the author has with EMF neutralization in whatever form it takes is that potentially the neutralizing frequency could make the patient sensitive to that frequency too, just like overexposures to certain foods, chemicals, etc., can cause

48

allergies. This concern is warranted in part by the English study's findings that, over time, the EMF sensitizing dose may need to be changed — in other words, it becomes a reactive frequency. Therefore, as suggested by one of the testers, several vials of neutralizing charged water, each charged at a different frequency, could be rotated as needed to minimize possible allergic reactions to just one vial.[10] Similar charged water experiments have also been done with ES patients at the Environmental Health Center in Dallas, Texas.

In the United States, diagnosis of ES patients was further demonstrated by the use of a binocular iriscorder to examine the patient's eyes while EMF exposed. This test is called a pupilography. The iriscorder device determines when and at what rate pupil dilation or constriction occurs. Comparing the changes in the eyes to EMF reactions in patients produced a direct correlation. When the ES patient was reacting to EMF reactive frequencies, the left and right eyes simultaneously dilated. At other frequencies pupil dilation was not synchronized. The doctors administering this test concluded that the EMFs were interfering with the body's autonomic nervous system at each individual's reactive frequencies. Study results showed neurological symptoms as dominant.[102]

Provocation testing must be carefully done to ensure other contaminants in the area are minimized. The room should be environmentally clean of chemical and electromagnetic co-factors that could interfere with test results.[4, 10, 102] Care should be taken that the patient is minimally exposed to reactive EMFs, as continued testing sensitizes the ES patient to more reactive EMFs, and diminishes the patient's health further. England's ES medical team notes that medical staff need to be present and prepared during EMF provocation testing.[10]

ES Legal and Economic Tangles

After the patient is diagnosed as electrically sensitive, legal and economic problems become apparent. In an ordinary illness, if the cause was work-related then worker's compensation would be pursued. If the illness was severely disabling, but not work-related, Social Security benefits would be pursued. A private disability insurance claim could be filed if available. But, ES is not yet a medically recognized illness in many ways and therefore traditional health/disability insurance payment, legal avenues, and federal/state benefit programs are difficult to access.

The chemically sensitive have had the same scientific/medical/political roadblocks as the electrically sensitive and have made some strides in accessing proper medical/legal/governmental help. If you are ES and MCS, the path may be easier, as these roads have been successfully crossed by other ES/MCS patients. Some MCS patients were able to get Social Security benefits after being diagnosed as having "environmental illness" or "immune system disregulation".[86] (See Medical Resources, page 72.) Check with Social Security regarding your individual case.

Social Security benefits can take up to two years to get, usually with a five month minimum wait. Of course, two years is a very long time for a seriously ill person to wait with no economic assistance in the meantime. The Lupus Foundation of America, in particular, has been pressing for reduction of the two-year Social Security and Medicare wait for its members under their health care reform agenda.[84] So, the environmentally ill are not the only ones waiting up to two years for benefits. Streamlining of this governmental process is certainly necessary.

ES needs to be specifically included under the umbrella of environmental illness for Social Security and other benefits in future public policy. Further, we, as ES patients, need insurance reimbursement under traditional medical insurance plans when we use an environmental medicine specialist. Currently, this is not happening and medical care is generally out-of-pocket no matter what insurance coverage the ES patient has.

Legal roads regarding product liability and injury are very new in the EMF area.

In 1994, the American Bar Association Journal reported that EMF cases were being filed at the rate of about one per month. These are primarily power line EMF cases; ES cases are coming but so far are not an established area. Worker's compensation needs are also not being well handled regarding ES cases. As ES becomes more well-known and accepted, benefits will eventually follow.

Carpal tunnel syndrome cases have had similar problems. According to the National Association of Working Women in their book VDT Syndrome, computer-related health problems generally have not been recognized as work-related. Reasons they cite include the variety of pain symptoms which are not outwardly apparent, gradual development and the possibility of many potential causes or co-factors. They also point to the tendency of doctors to avoid risking their professional reputation where scientific data is lacking.[93] CTS cases now are many, with some successful results. (See Legal Resources, page 73.)

Now that the Americans with Disabilities Act is law, environmental illness patients are testing the limits of it. ES patients are faced with electronic barriers to employment. Also, EMF barriers restrict shopping and social outings too. If a person cannot access the workplace due to ES, that person has a disability. Maybe they can access another workplace instead; maybe they cannot. They might be able to work only from home; they might not be able to work at all.

Electrical Sensitivity Treatment Options

Is there a cure for ES? If you find one, please let us all know! Presently, there are ways to improve your health and reduce your symptoms, but there is not one clear path for everyone besides avoiding those things that bother you.

WARNING: The following treatment options are ways some chemically sensitive and/or electrically sensitive patients have sought relief of symptoms. You must decide for yourself what course to take, if any; medical guidance in making your decision is highly advisable.

1.   <u>Improving the Environmental Quality of Your Home</u>

Avoidance of electrical/EMF sources as much as possible is the most satisfactory means of reducing ES symptoms. (See EMF Reduction in the Home, page 67.)

Other unhealthy factors beyond EMFs in the home entail chemical emissions from synthetic materials in plastics, carpeting, furniture, drapery, clothing, bedding, cleaning products, building construction materials, etc. MCS patients often find that they need to minimize all synthetic materials in their home and have basic, untreated natural materials — cotton, wood, porcelain tile, etc. — instead.[49, 78, 86, 94, 106] (See Information Resources, page 72, for guidance in your individual situation.)

Having an environmental illness often requires a person to become housebound or have restricted activities beyond the home due to their inability to control environmental exposures beyond their home.

If the home is unlivable/unhealthy for the environmentally ill person, then the patient feels that they have *no* safe place to be away from chronic chemical/EMF exposures. This is very stressful and frustrating.

A perfect example of this predicament recently occurred in the United States government's new and first low-income housing for the environmentally ill, Ecology House in California. Soon after the MCS tenants moved in, they became more ill reacting to the building's unhealthy construction materials. EMF problems may be an unsuspected factor in the Ecology House, too. These already ill people are further ill from their new home. Several Ecology House residents have had to sleep outside, sleep in the bathroom, breathe from an oxygen tank, or go to a friend's house to escape the unhealthy environment within their own home.[98, 116]

ES patients have similar problems. Many ES and MCS patients have resorted to camping out in the wilds to avoid both electrical and chemical exposures to the utmost. Only then do they feel better.

Sorting out which factors are bothering you is very helpful on the road toward feeling better again. Other sensitivities as noted in Diagnosis of Electrical Sensitivity, page 37, should also be considered.

2.   Natural Fiber Clothing

Clothing made of synthetic fibers (polyester, acrylic, rayon, etc.) adversely affect many of the chemically sensitive.   Undyed, untreated, organic cotton clothing is preferable for MCS patients.

ES patients need to worry about the electrical, rather than the chemical, properties of clothing fibers. Natural fibers such as cotton, linen, and silk are better for the ES patient too.

Synthetic fibers in general create static electricity problems, may act as an EMF wave-guide material, attract positive ions, hold heat and are best avoided.[51, 97]   For example, a woman with a burning skin reaction from computers in the office may find wearing nylons intolerable at work. Nylon, electrically, is an insulator. When EMF exposure creates induced currents in the body and produces heat, nylon prevents adequate release of the induced heat, intensifying the reaction.

3.   Metal Avoidance

Metal jewelry, watches, and metal enclosures can be troublesome for some ES patients. Avoiding these items may be necessary.[9, 10, 105] (See Metal Sensitivity, page 42, and Reactions to Watches, page 43.)

4.   Grounding

When you walk across a synthetic carpet and reach for the metal doorknob, sometimes you get that familiar electric shock. That shock discharges the energy stored by the friction of walking across the floor.

Storing electricity in the body can build up stress. To release stored electricity, walking barefoot on the bare ground or grass is helpful. Distancing yourself from power lines and other EMF sources while doing this grounding technique is best. This method is called grounding from the electronics term relating to discharging electricity to the earth.

The Environmental Health Center in Dallas has recommended this grounding technique to ES patients. Twenty minutes per day of grounding has helped some ES patients. An alternate method is to lie down on the ground if the weather does not permit bare feet.

5.   Product Warning

Beware of all products claiming to reduce your EMF exposure or reduce the effects of EMF exposure. The author has tested some of these only to find that they did not stop her ES symptoms from occurring when EMF exposed.

An important point to keep in mind is that if the gaussmeter can still measure EMFs, the product is not 100% effective in protecting you. Induction heating would still be occurring.

All EMF protection products should be considered strictly experimental. If you choose to test any of these yourself, check the refund/return policy of the dealer first.

Some products are emitting a field of their own which is claimed to counter other EMFs. ES patients would do best to avoid all EMF-producing products of any sort. Science needs to understand that the best way to reduce EMF reactions is not by using one EMF to cancel/fight another EMF, but all EMFs need to be reduced as much as possible. Metal EMF shielding technology implanted at the time of product design is a necessity needed now on all electrical appliances.

Magnets have an energy field also. Any magnet therapy should also be considered *questionable and experimental* for the ES patient, particularly when placed near the head.[105]

6.   Antigen Therapy

A common method of treating environmental illness patients in general is first by allergy testing of foods, chemicals, and sometimes metals. Once allergic substances are pinpointed, those substances can be avoided. Also, chemical antigen shots may be administered representing neutralizing doses for those allergens. EMF reactions have been neutralized chemically, but documented information in this area is scant.[4, 10]

While chemical antigens have provided some reported symptom relief for food and chemical "allergy" symptoms, daily shots are usually required to maintain the program.

A question arises regarding the long-term use of antigens made from your reactive foods/chemicals. The body produces antibodies upon antigen injection for fighting symptom reactions.

7.   Charged Water Technique

Another allergy test involves determining the ES patient's reactive EMF frequencies and non-reactive frequencies. The focus is on finding at least one non-reactive frequency. (See Diagnosis of Electrical Sensitivity, page 37.)

For EMF "allergy" treatment, vials of water are charged at a specific non-reactive EMF frequency or frequencies. The water used is saline or mineral water that the patient does not react to. The ES patient is told by the doctor to grasp the vial of charged water for five minutes once or twice daily to reduce general EMF sensitivities. Also, they can hold a charged water vial to reduce ES symptom reactions from EMF exposure.[4, 10]

Once non-reactive EMF frequencies are determined, future vials can be charged without the patient present. Then, the vials can be mailed to the patient without need of a doctor's visit each time.

After using a charged-water vial to neutralize a severe EMF reaction, ES patients may feel the effects of that vial discontinued — drained. Charged vials lasted about two weeks in the United States while England's water held the EMF charge for about one month or more.[10]

EMF neutralizing vials of charged water representing individualized non-reactive EMF frequencies has been done by Cyril W. Smith, Ph.D., in cooperation with Dr. Jean A. Monro at the Breakspear Hospital in England. The Environmental Health Center in Dallas, Texas, has also used this technique with Dr. William J. Rea's cooperation.

This technique may provide a way for the patient to neutralize an EMF reaction; however, the initial EMF testing session can provoke all symptoms. EMF testing may also elevate sensitivities, particularly if a reactive EMF frequency is held for more than 20 seconds. Charged vials could create a dependence on certain EMF frequencies to the point that those frequencies also produce "allergic" reactions. At that time renewed testing is needed to reveal current neutralizing frequencies.[10]

8.   Salt Water Bath

Some ES patients report that bathing in salt water, either a salt water bath (using rock salt) or in the ocean, is helpful to temporarily alleviate their symptoms.[10] The reason could be because the salt water is electrically conductive and helps absorb the ES patient's own EMF signals. (See Metal Sensitivity, page 42.) Ordinary table salt includes additives, including an aluminum derivative, so is best avoided, even for bathing.[106]

9.   Detoxification

If toxicity from chemical or metal exposures is present, detoxification can reduce stored toxins in the body.[59] By detoxification, the body may reduce EMF sensitivity as well as reduce chemical sensitivity. "Detox" includes exercise and saunas to sweat out toxins. Home saunas can be a problematic EMF source, so turning the sauna off after preheating it may work better for ES patients.

At the Environmental Health Center in Dallas, detoxification has included medically supervised saunas with vitamin B_3 (niacin) supplements and exercise to draw out poisons through the skin. This therapy is called the Hubbard Technique.[86] Other vitamin and mineral supplements are medically adjusted to compensate for the body's loss of these during sweating.

Also, detox means not adding back poisons that are being excreted. One way to reduce toxic intake is by using fresh, organically grown foods — no pesticides, herbicides, fungicides, or chemical fertilizers. Also healthy is avoiding packaged foods which contain artificial ingredients such as sweeteners, dyes, colorings, flavorings, pesticides, flavor enhancers, preservatives, etc. Food additives have been implicated in some cases of attention deficit disorder (ADD) in hyperactive children.[58] The Benjamin F. Feingold, M.D. diet has helped some hyperactive children return to normal health. A basic natural foods diet like Feingold's can determine whether food additives are a factor in ADD or in a person's MCS/ES symptoms.

An elimination diet — eliminating one food at a time — could be used to check whether specific foods are producing allergic reactions. By keeping a list of your foods eaten, you can assess whether a particular food is causing a reaction if eaten again. Varying your meals so that you eat a good variety of species and types of food helps reduce food allergy reactions caused by eating one food too often. Three of the most common food allergens are milk, wheat, and corn. Using variety in your diet may mean switching to other grains such as spelt, quinoa, and amaranth.[47, 48]

Herbs can also play a part in the detoxification process to help cleanse the liver, colon, kidneys, and blood of stored toxins.[107, 108] The liver is the body's main detoxification organ.[61, 124] A naturopath who understands herbs may be helpful if you are interested in herbal therapy. MCS patients may not be able to use some or all herbs, particularly if the herbs are not organic.

Periodic fasting is an old form of detoxification. Patients who are already weak or malnourished should not attempt fasting as it will further weaken them. Fasting can help remove stored toxins, particularly if lots of pure water is taken. Chemically sensitive patients often develop intensified symptoms when hungry or fasting, because stored toxins are resurfacing in the blood.

Detoxification can be very difficult for the chemically sensitive. Care needs to be taken that "detox" is done gradually, and preferably medically-supervised to minimize the reactions and ill health possible when the stored toxins further mobilize in the body. The detox method can seem like you feel worse before you feel better.

An intensive doctor-assisted detoxification program may be warranted when all else fails to help the MCS/ES patient. The Hubbard Technique is one example. Another interesting example is Gerson Therapy, carried out at a Mexican hospital, and used primarily for advanced-stage cancer patients. This "detox" and rebuilding therapy primarily consists of 13 glasses of fresh, raw organic fruit and vegetable juices per day, three organic vegetarian meals daily, enemas to speed detoxification, pancreatic enzymes and selected vitamin/mineral options individualized to the patient's needs. This therapy is outlined in Max Gerson, M.D.'s, book, <u>A Cancer Therapy</u>.[60, 61] Other healing and detoxification programs have also used raw organic vegetable and fruit juices with a natural diet. The Bircher-Benner clinic, Zurich, Switzerland begun in 1897 is another example.[37] One drawback is that using an electric vegetable juicer is going to be off-limits for most ES patients without help.

Detoxification of any kind is assisted by exercise. Aerobic exercise like bicycling, jogging, swimming, etc., can be a great stress reliever, too. Many factors determine our health — quality of our diet, exercise, air, water, mental attitude, etc.

10. <u>Energy Balancing</u>

The body is electromagnetic. When a person is ES, their adaptability to other EMFs is impaired.

Toxic chemical exposure, metal toxicity, and prolonged EMF exposure are all possible ways the body could become electromagnetically impaired. Each type of chemical and metal is electromagnetically unique. A preponderance of certain chemical or metal deposits in the body could change the way that the body reacts electromagnetically with its surroundings. Nervous system and immune system damage from toxic exposures may also impair the body's EMF tolerance capabilities.

The philosophy of the acupuncturist is that the body manifests energy imbalances before physical symptoms appear. By correcting energy imbalances early, disease processes are diverted before symptoms manifest, thereby averting disease at its earliest stage.[73]

A medical example of the promise for electromagnetic interpretation of disease comes from medical experiments at Yale University in the 1970's. Ninety-eight percent of the female cancer patients studied showed a <u>negative</u> electrical potential at the cervical cancer site relative to the abdomen. When female non-cancer patients were examined, a <u>positive</u> electrical potential was measured 81.9% of the time instead. The conclusion proposed from this study was that cancer is the result of changes in the body's organizational electromagnetic fields.[42]

Apparently health and disease each have their own electricity. There is also evidence that every disease has an electromagnetically unique pattern.[85] Finding the electricity of health and reducing the electricity of disease in whatever form (chemicals, metals, electromagnetic exposure, etc.) appears to be the challenge of the twenty-first century.

Perhaps in the future a mere blood sample checked by spectrum analysis for absorption rates at various EMF frequencies could disclose foreign bacteria, viruses, fungi, chemicals, metals, etc., invading the body. A check for unhealthy EMF frequency absorption rates could indicate the problem and lead to a custom-designed remedy suited to the patient's EMF picture. A cataloging of homeopathic remedies by their EMF absorption rates could provide a ready remedy electromagnetically compatible with that person.

The use of EMFs in the treatment of illness by conventional medicine is in the early discovery stages. This trend is called energy medicine or electromedicine.

Doctors at Scripps Clinic in La Jolla, California, recently used a battery-operated EMF device to induce sleep in insomniacs. By placing this device on the patient's tongue for 20 minutes before bedtime, sleep was caused in 80% of the insomnia sufferers. This EMF gadget delivers pulses of 27 MHz (27,000,000 Hz) EMFs to the brain.[80] Therefore, brain waves are being affected. Notice how EMFs induce sleep in this case and with some ES patients too.[102] Some people can get insomnia from EMFs instead; both are neurological changes that can be caused by EMFs.[39]

ES patients should avoid such EMF-treatment devices which may only make them worse. These EMF-emitting devices are experimental and highly questionable in light of the potential health hazards of EMFs, particularly cancer. (See The EMF Cancer Issue, page 62.)

Many forms of alternative healing methods rely on the balancing of the body's "vital energy" as the Chinese call it. Acupuncture is an ancient technique used to balance the body's electromagnetic fields. Other energy balancing forms are acupressure, homeopathy, Therapeutic Touch, Tai Chi, and Qi Gong.[56, 74, 76, 105, 111] These alternative therapies are designed to help the body balance its own electromagnetic field energy without artificial EMF sources.

Some ES patients and allergy patients claim to have been helped by various energy balancing therapy.[105, 111] How you proceed is your own decision — what helps one person will not necessarily help another person, as each person is electrochemically different. It would seem to make sense for detoxification to accompany any energy balancing where chemical or metal exposure is a suspected ES initiator, so that the corrections would be more lasting.

11. Miscellaneous

A few other possible factors that may determine the severity of ES reactions include antioxidants, minerals, acid/alkaline balance (pH), and enzymes.

Antioxidants are a group of vitamins, minerals, etc., that are known to reduce cellular damage from radiation and other degenerative processes. Antioxidants work by diminishing free radical activity caused by toxic chemical and metal exposures, EMFs, and stress.[108] Common antioxidants are vitamins C, E, and the mineral selenium. Some herbs such as chaparral also have antioxidant properties. Superoxide dismutase is an enzyme with antioxidant properties.[24, 52]

Whether antioxidants are helpful in delaying or diminishing ES symptoms from EMF exposure is not yet clear.

Electromagnetic field exposure has been shown to displace minerals, particularly calcium. Minerals are the body's metals. All minerals in the body are called electrolytes and have a small electrical charge.[28] Minerals help in the transmission of nerve signals through the body. When an imbalance of minerals is present, then the body's electricity may become imbalanced too.

Dr. Max Gerson found that the body's potassium is inactive relative to sodium in a chronic disease state.[61] A September, 1991 Journal of Hypertension article (p. 167) parallels that concept by stating how a potassium deficiency can have an adverse effect on the body's nervous system. The article cites the sodium-potassium pump on the nerve cell membrane as the problem location. As previously noted, neurological problems are quite frequently noted in ES patients.

The place for minerals in the ES picture needs research.

Another factor needing further investigation relative to the ES picture is the body's acid/alkaline balance (pH). Blood is normally slightly alkaline. Maintaining an alkaline basis in the blood may reduce microorganism infestations, as most microorganisms cannot live in a high alkaline environment.[43]

Some ES patients have experienced increased symptoms when taking acid-forming foods such as alcohol and sugar. Eliminating these products is recommended in general for environmental illness patients, according to Sherry Rogers, M.D., an environmental medicine physician.[59]

Alkaline-forming foods are primarily vegetables and fruits.

Enzymes help the body process foreign material to detoxify the body. MCS patients are often low in some enzyme levels, perhaps due to chemical overload. Enzyme function and enhancement is another research question regarding ES symptoms and their diminishment [94, 106]

12. Drug Intervention

Gamma-hydroxybutyrate (GHB) has been used experimentally in the treatment of ES symptom reduction.[134] Currently, GHB treatment is not available in the United States because GHB cannot be legally sold here. GHB has known central nervous system effects including insomnia relief and muscle relaxation. Apparently, it is because of these neurological effects that GHB has been shown to temporarily reduce some ES symptoms that have a neurological basis. The relief from GHB is a form of symptom suppression, not a cure. Daily pills are required to maintain the effects. However, the outcome of using a muscle-relaxant for long periods of time is unknown. This method should be considered experimental.

What are Electromagnetic Fields (EMFs)?

Electromagnetic fields (EMF or EMFs) are fields/waves of energy that are emitted by all electrical sources. EMFs are also produced by the earth and natural weather conditions. Common electrical sources of EMFs are electric power lines, electric home and office appliances, motors, wall wiring, electrical substations, transformers, and radio/radar/microwave transmitters. These sources create non-ionizing radiation — radiation which is not currently known to break molecular bonds like x-rays and ultraviolet (ionizing radiation) do.

The sequence of EMFs from low-energy to high-energy is displayed in the electromagnetic spectrum diagram which follows, showing all of the wave forms relative to each other, based on their alternating current (AC) cycles per second. Extremely low frequency (ELF) sources are all electrical utility dependent products. Very low frequency (VLF) products include computer monitors and television sets. Direct current (DC) (no alternating cycles) occurs at the zero point on the chart. Some ES patients may be bothered by DC, but generally DC is much less troublesome than AC.

The more the alternating current cycles per second, the more energy in the EMF wave. Therefore, x-rays have more energy than most EMFs while power-line type radiation (ELF) has the least energy.

Science has been debating the health hazards of non-ionizing radiation for over twenty years. The more EMFs have been studied, the more health hazards are becoming apparent.

Currently, low-frequency EMFs are being investigated primarily in connection with Alzheimer's disease and cancer.[87, 95, 113, 128] Other likely health hazards of low-frequency EMFs are miscarriage, birth defects, and electrical sensitivity.[2, 53, 68, 69]

62

ELECTROMAGNETIC SPECTRUM

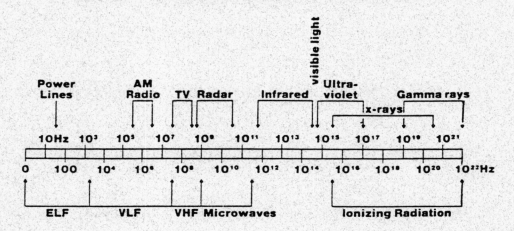

Source: United States Department of Energy[127]

Hz = Hertz, meaning cycles per second

10^3 = 10 x 10 x 10 = 1,000 cycles per second

ES patients can have EMF frequency sensitivity problems at any frequency. Inexpensive gaussmeters cannot measure most frequencies in the VLF range. Inexpensive equipment for measuring radio frequency and below in the home would be an interesting and useful item, but is not currently commercially available. When these instruments become available, ES patients will more easily find and "see" fields they may be reacting to.

The EMF Cancer Issue

Do electromagnetic fields (EMFs) like those found near electrical power lines and appliances cause cancer? Scientists in over 22 countries are investigating the possible EMF/cancer link, with more than 200 health-related EMF studies now underway worldwide.[128]

Evolution of the cancer issue has brought a focus to certain cancers that appear frequently in EMF studies: leukemia, breast cancer, and brain cancer.[128] Studies of children living near high-voltage power lines and workers exposed to higher-than-normal EMFs (electricians, utility linemen, etc.) both show increases in leukemia. Electrical workers in the United States and Sweden both develop a higher-than-average rate of brain cancer.[22, 77]

Another aspect of the current EMF debate involves the sometimes noted increase in breast cancer incidence among *both* men and women in high electrical exposure occupations. The National Cancer Institute's statistics indicate that 1 out of every 8 women in the United States will get breast cancer. Power line EMFs in some studies reduce levels of the hormone melatonin, a natural tumor suppressor. By reducing melatonin, EMFs are suspected of promoting and perhaps initiating the cancer process.[112, 128]

In the National Cancer Institute's June 1994 Journal, a University of North Carolina study found that female electrical workers had above average breast cancer deaths. The electrical workers' exposures to power line EMFs as telephone repairers, installers and line workers is suspected as a factor in their increased breast cancer incidence.

Several prior studies of male electrical workers also found higher-than-normal breast cancer rates for them. For example, in 1990 the Hutchinson Cancer Research Institute in Seattle, Washington studied 250 male breast cancer patients. They found that workers with the most EMF exposure — electricians, power workers, and utility linemen — had six times the level of breast cancer as workers without high EMF exposure.[77]

According to the United States Department of Energy's new booklet, Electric Power Lines, "Laboratory studies have shown that it is unlikely that EMF can initiate the cancer process. Some studies suggest, however, that power-frequency EMF may promote development of certain existing cancers."[128] Studies now on-going are testing both the ability of EMFs to cause cancer promotion and cancer initiation.

Many toxic chemicals are known or suspected cancer initiators. A 1992 study by Frank Falck, M.D., at the University of Connecticut showed high levels of the chemicals PCB, DDT, and DDE in cancerous breast tumors compared with benign breast lumps. A subsequent study at Mount Sinai School of Medicine, New York, found four times the level of breast cancer risk for women with the highest DDE blood levels.[44] A United States study of 200 chemically sensitive patients found DDT and DDE pesticides in 62% of patient blood samples. These pesticides occurred more often than any other type of pesticide found.[94] Chemical exposures combined with EMF exposures sound like a recipe for cancer.

Prudently, the Swedish National Electric Safety Board in 1993 announced that "...the Board will act on the assumption that there is a connection between power frequency magnetic fields and childhood cancer, when preparing regulation on electrical installations."[128]

Electromagnetic fields are not only found near power lines, but also emanate from home and office electrical appliances. Computer monitors, fluorescent lights, microwave ovens, and electric clocks are common EMF sources. At work, computer users are exposed to EMFs for prolonged periods of time.

The time has come to plan and implement preventative measures that reduce EMF exposure throughout our electrical society, indoors and out. The potential health and legal implications of not doing so are immense.

Warning Signs of EMF Problems

Certain signs of possible EMF exposure situations can be noted without having a meter, merely by looking around. (Or for an ES person, just by being there.)

An environmental problem should be suspected when an ES patient repeatedly reacts to a particular location. The question needing to be solved is specifically what the person is reacting to. Keeping chemical exposure possibilities and other sensitivities in mind, EMF sources need close analysis.

Clues of EMF exposure can be detected by static on radio station channels or distortions apparent in the television set picture or computer monitor screen. Sources of EMF-caused interference include power line radiation, appliances, and wall wiring electrical problems.

Locating the source of all EMFs is necessary for ES patients in order to plan EMF reduction strategies for minimizing symptoms.

A look outside may find EMF sources such as electric power lines, power line transformers (metal cans on the utility poles or metal boxes on the ground), an electrical substation, and radio/radar/microwave transmitting towers. All of these can be troublesome for the ES patient.

After a thorough check for potential EMF sources, measuring EMF emissions is the next step. Either using an EMF testing service or obtaining the measuring equipment and testing EMFs yourself will help you "see" what levels of EMFs are present, where the highest readings are, and where hidden EMF sources are. (See EMF Resources, page 74.)

What a Gaussmeter Will and Will Not Tell You

A gaussmeter is a measuring device to detect low-frequency magnetic fields. Electromagnetic fields have two components — an electric field and a magnetic field. An ELF (Extremely Low Frequency) gaussmeter will measure the 60 Hz (United States) magnetic field from power lines, household wiring, and some electrical appliances. The ELF magnetic field is currently of most concern in ongoing scientific cancer studies regarding EMF health effects. An ELF gaussmeter is the type of equipment most commonly used for home EMF measurements.

However, ES patients can be troubled by both electric and magnetic fields at any frequency. Therefore, both ELF electric and magnetic field readings ideally would be taken at the home of the ES patient. An ELF electric field meter measures the 60 Hz electric part of the EMF. EMFs in the ELF through microwave range are generally most limiting and troublesome in the daily lives of ES patients.

Measuring higher frequency magnetic fields, such as very low frequency (VLF) from computer monitors requires a VLF gaussmeter. Inexpensive ELF gaussmeters are often inaccurate for measuring computer monitors and other appliances that also produce these higher frequency fields. VLF fields record high readings in ELF gaussmeters not sophisticated enough to differentiate between ELF and VLF magnetic fields. Before purchasing a meter, first ask meter manufacturers what field(s) their meter is designed to measure and what (power lines, computers, etc.) you could accurately measure with the meter.

1.　　Gaussmeters generally do not measure electric fields (E-fields), only magnetic fields (H-fields) from electrical sources. Some gaussmeters have an electric field measuring component in addition to the magnetic one. Power line and electrical appliance emissions are 60 Hertz electric and magnetic fields. You would need an ELF electric field meter or a gaussmeter with an electric field measuring feature to measure the electric component of the ELF field. (See diagram below.)

E-Field	ELF Electric Field	VLF Electric Field
H-Field	ELF Magnetic Field	VLF Magnetic Field

2.　　Gaussmeters do not measure all forms of electromagnetic radiation, only low frequency magnetic fields. Sound waves, x-rays, radon, microwaves, etc., require other types of detection equipment instead of gaussmeters.

3.　　Gaussmeters do not read magnetic field levels below their own level of sensitivity. For ELF gaussmeters, a sensitivity level down to .1 milligauss is preferable; for VLF gaussmeters, .01 milligauss. The magnetic field reduces with distance from the source, but does not reach zero when your meter does — the field reduces to infinity. Someone sensitive to these magnetic fields may become ill at levels below those you can measure. (They may also be reacting to other types of fields — sound, electric field, microwaves, radio wave harmonics, etc.)

4.　　Most inexpensive gaussmeters will only measure one direction of the field at a time. To find the highest reading, turn the meter horizontal, vertical, and sideways to the EMF source. A triple-axis meter will measure the three dimensions at once for one combined milligauss reading.

5.　　Gaussmeters show you where measurable magnetic fields are and their measurement. However, there are no United States government health regulations regarding power line radiation.

EMF Reduction in the Home

Time Magazine, October 26, 1992, reported on a milestone Swedish study linking extremely low frequency (ELF) electricity to cancer. The study, conducted by Stockholm's Karolinska Institute, evaluated cancer incidence among almost 500,000 people who lived within 328 yards of high-voltage electrical lines from 1960 to 1985. They found a direct correlation between childhood cancer incidence and incremental increases of ELF magnetic field exposures. Cancer in children was lowest below 1 milligauss exposure. Cancer rates were three times higher where 2 milligauss levels occurred and four times higher where three milligauss levels occurred. Leukemia was the cancer type noted in the study.[22]

The conservative consensus building now is that ELF magnetic field exposure levels above 2 milligauss are best avoided by the public at large.[12] United States government guidelines or regulations have not yet entered this area. Of note, the Bonneville Power Administration in Portland, Oregon — an agency of the Federal government — discourages access of their electrical power line right-of-ways for new uses (playgrounds, parks, etc.) which would increase public EMF exposure.[128]

For electrical sensitivity patients, EMF exposure should be as low as possible, given the economic feasibility of doing so.

Having your home tested for EMFs may uncover some surprises. One often noted problem that is fairly easy to correct is improperly grounded electrical wiring in the home. Published accounts of ES patients using EMF testing technicians point to wiring problems from ungrounded electricity as a common item that can be resolved, reducing EMF exposure.[65, 109]

An EMF testing survey entails measuring power line radiation, wall wiring, and appliance EMF emissions. Power line radiation coming into the home can be detected by shutting off all electricity in the house at the electrical fuse box. Then, power line radiation coming indoors can be measured with ELF testing equipment.[104]

If you are having a problem with power line EMFs, ask your EMF testing technician to assist by acting as a negotiator when discussing the power line problem with your electric utility. Power line remedies are, however, usually difficult to get done, may prove expensive to the customer, and may not be helpful enough for the ES patient.

While EMF testing is useful, it is important to keep in mind that people who are electrically sensitive may feel reactions to EMF levels below those easily measured. For instance, some severely sensitive patients report reactions to power lines or airport radar one mile or more away. Only very sensitive meters could detect power lines that far away.

Being ES means that you are hypersensitive to EMFs, even small amounts. While your electric utility company and your independent EMF testing technician may say that your home has very low EMF levels, for you that may not be enough. This is how many environmental illness people become homeless, of sorts. Moving may be your best option in this case, but the question is "Where can you safely move to?"

Checking your home for EMF wiring and grounding problems is done by measuring the ELF field inside with all of the lights turned on. A map of the home's floor plan may be drawn and EMF readings recorded at various points. Levels of ELF magnetic fields recorded above three milligauss near walls without appliance or power line intervention are commonly miswired circuits out of compliance with the National Electrical Code (NEC).[104] All changes made to grounding and wiring must comply with NEC and any local building code to minimize fire hazards and be legal with current building regulations. An electrician can help with needed corrections.

Locations in which you spend more time should be thoroughly checked. Appliance emissions could be tested too, but for the ES patient, using or being near operating electrical appliances is best minimized anyway.

In addition to EMF testing, some EMF technicians provide EMF reduction services and make suggestions as to how you can reduce EMF exposures. Some electric utility companies now measure ELF magnetic fields inside and outside the home as a free service. However, they usually do not give advice for home EMF reductions.

The four primary ways to reduce EMF exposure are by 1) increasing your distance from the EMF source(s), 2) reducing time spent in unavoidable EMF areas, 3) using appliances with no or reduced EMF emissions, and 4) shielding, when economically feasible.[128]

ELF magnetic fields easily penetrate walls, so power line radiation is penetrating, even indoors. These magnetic fields are difficult to shield, usually requiring a special nickel-iron alloy of the Mumetal variety or low-carbon steel. Generally, the more nickel content in the alloy and the thicker the metal sheet, the more shielding is provided. Mumetal is helpful for reducing small EMF source problems, but expensive and cumbersome for large products.[12, 35] EMF technicians sometimes have magnetic shielding material for EMF reduction jobs. (See EMF Resources, page 74.) ELF electric fields are easily shielded by electrically grounded metals such as copper or aluminum. The electric fields are less penetrating than the magnetic fields and are reduced somewhat by walls and other obstructions.

The most ill ES patients turn off electricity to all or a part of their house to reduce ES reactions and to sleep. Essential equipment needs to be accounted for during the electrical "down-time". Refrigerators cannot be safely turned off for more than one hour, unless other arrangements are made for food contents — like using an ice box, etc. Electric clocks will need to be reset after the down-time, so battery-operated ones are preferable, unless the ES person is troubled by battery sources as well. Computer data needs to be saved prior to down-time, if you or someone else in the home is still using a computer.

How do you feel when all electricity in the home is shut off? Some ES patients find significant help using this technique — particularly for sleeping at night after turning the electric circuit to the bedroom off. This is a good way to check whether other factors may be bothering you also.

When a patient is both MCS and ES, a serious question develops regarding what utilities they can use to cook with and to heat their home — natural gas or electricity? Usually electricity is used for MCS patients, because many are very sensitive to the natural gas. Once electricity becomes an unhealthy factor for them, some turn off electricity to part or all of their home in order to minimize ES reactions. Doing daily living tasks then becomes very difficult or impossible without help.

Ways to reduce EMF exposure for the ES patient include unplugging all unused appliances, discontinued use of the television set and computer, using gas (if not MCS) or other substitute appliances, and using incandescent light bulbs but not energy-efficient or fluorescent bulbs.

Some appliances have EMF readings when plugged in, but turned off, so unplugging unused appliances is most helpful for EMF reduction.

ES patients may become sensitive to a television set at 40 feet or more, even one turned on next door.[9] So, ES patients often not only do not watch television from a distance, they do not have one operating at all in the house. Television EMFs easily penetrate walls and are difficult to reduce. For the ES, reduction is usually not enough.

Gas appliances are sometimes helpful for ES patients that are not MCS. Gas stoves, refrigerators, heaters and ovens can reduce EMF exposure. Older gas appliances have a pilot light that burns a small amount of gas fuel constantly which may increase or lead to chemical sensitivity problems, although use no electricity. Newer gas models generally have an electronic feature that only activates the gas when you turn the appliance on. This electronic part is electrical, thus an EMF source. Newer gas appliances may need to be EMF tested prior to purchase to determine whether they suit your situation.

The most severe ES patients cannot use a standard phone, due to the EMF exposure from the phone earpiece. Some United States patients are using a speaker phone to remedy this problem. An airhose phone adapter is available in Sweden to allow ES patients to use a phone at a distance of six feet away.

Fluorescent lighting is a common irritant to ES patients. Unfortunately, fluorescent lighting is economical relative to other lighting and is therefore in most commercial and governmental buildings. Doctor visits may uncover fluorescent lighting there too. Environmental medicine doctors have found it necessary to turn off fluorescent lighting in the patient area when ES patients are there. Use of daylight from windows and floor lamps with incandescent bulbs instead may avoid or reduce EMF reactions that could occur with lighting from fluorescent bulbs.[92]

Other innovative solutions to EMF reduction problems reported by severely sensitive ES people include grounding metal window and door frames, removing non-essential fuses from the car (clock, etc.), disconnecting built-in electric clocks, re-wiring the house with Teflon-coated wire, re-wiring lamps with metal-shielded wire, and turning the knob in the refrigerator to "off" to prevent the motor from running while in the kitchen.

Endnote

This book can in no way describe the broken dreams, the failed marriages, lost careers and homes, shattered family relations, ridicule, and hopelessness that haunt the environmentally ill.

Their voices echo similar words, even when they know no others likewise affected. When the Europeans explain their own symptoms and so describe us, we feel whole again — validated.

And, by finding others with familiar problems, we find friendship with those who understand us without explanation. After the difficult journey, it is good to find a friend. Though we may not be able to help each other get well, we discover that we are not alone.

RESOURCE GUIDE

Information Resources:

- National Center for Environmental Health Strategies, 1100 Rural Ave., Voorhees NJ 08043, Phone: (609) 429-5358.

  Chemical sensitivity information and newsletter <u>The Delicate Balance</u>.

- The Environmental Health Network, P.O. Box 1155, Larkspur CA 94977, Phone: (415) 541-5075.

  Support group network for chemically sensitive with newsletter — <u>The New Reactor</u>.

- Human Ecology Action League (HEAL), P.O. Box 49126, Atlanta GA 30359, Phone: (404) 248-1898.

  Chemical sensitivity information and newsletter <u>The Human Ecologist</u>.

- Electrical Sensitivity Network, c/o Weldon Publishing, P.O. Box 4146, Prescott AZ 86302.

  United States electrical sensitivity group in the formative stages. Contact the publisher for current group information.

- Association for the Electrically and VDT Injured, P.O. Box 15126, 10465 Stockholm, Sweden, Phone: (011) 46 8 7129065. INTERNET (worldwide web address): http://www.isy.liu.se/~tegen/febost.html

  International contact for electrical sensitivity groups worldwide. Their INTERNET information is excellent and available through many public libraries.

Medical Resources:

- American Academy of Environmental Medicine, 4510 W. 89th St., Prairie Village KS 66027, Phone: (913) 642-6062.

Environmentally ill patients who seek medical doctors specializing in environmental medicine can get referrals in their area.

- Environmental Health Center, 8345 Walnut Hill Ln., Suite 205, Dallas TX 75231, Phone: (214) 368-4132, Fax: (214) 691-8432.

William J. Rea, M.D. is the chief physician at this medical clinic for the environmentally ill. A pioneer in the experimental treatment of ES patients with the charged water technique.

- Allergy and Environmental Medicine, Ltd., Breakspear Hospital, Belswains Lane, Hemel Hempstead, Herts HP3 9HP, England, Phone: 01442-61333, Fax: 01442-66388.

Dr. Jean Monro is the medical director. English pioneer of the charged water technique for ES patients.

- Gerson Institute, P.O. Box 430, Bonita CA 91908, Phone: (619) 472-7450.

Information center for the Gerson Therapy, an intensive detoxification program. Primarily used for alternative cancer treatment and not environmental illness, but may be useful where other methods have failed and chemical detoxification is needed.

- Check your local phone directory.

Legal Resources:

- Robert Strom Foundation
  Michael Withey
  Schroeter, Goldmark & Bender
  810 Third Avenue, Suite 500
  Seattle WA 98104
  Phone: (206) 622-8000

Current central contact for EMF-related legal cases. Referrals to EMF lawyers in your area.

- Check your local phone directory.

EMF Resources:

- National Electromagnetic Field Technicians Association (NEFTA), 628-B Library Place, Evanston IL 60201, Phone: (708) 475-3696.

   Organization of independent EMF testing technicians that provides referrals to members in your area — United States and Canada. Some have metal EMF shielding material.

- VDT News, P.O. Box 1799, Grand Central Station, New York NY 10163, Phone: (212) 517-2802, Fax: (212) 734-0316.

   Provides publications VDT News and Microwave News regarding the health hazards of EMFs. List of gaussmeter manufacturers available for $1. Back issue reprint packets on current EMF topics, EMF Resource Directory, etc. Sample issue of VDT News, $5.

- National EMR Alliance, 410 West 53rd St., Suite 402, New York NY 10019, Phone: (212) 554-4073, Fax: (212) 977-5541.

   Publication Network News provides current information on EMF activist efforts, general EMF news, and has adopted the electrical sensitivity issue as an on-going feature in the newsletter.

- The Labor Institute, 853 Broadway, Room 2014, New York NY 10003, Phone: (212) 674-3322.

   Informative booklets entitled Electromagnetic Fields (EMFs) in the Modern Office and Multiple Chemical Sensitivities at Work. Other booklets and videos also available from this labor union group concerning worker health.

- University of California, Labor Occupational Health Program, School of Public Health, 2515 Channing Way, Berkeley CA 94720, Phone: (510) 642-5507, Fax: (510) 643-5698.

The Labor Occupational Health Program has a library with a special VDT collection regarding health issues of computer usage.

- National Association of Working Women (9 to 5), 1224 Huron Rd., Cleveland OH 44115, Phone: (800) 522-0925, or (216) 566-9308.

  Toll-free number is their Job Problem Hotline. Fact Sheet on fluorescent lights and health.

- Your local electrical utility company.

  Many electrical utilities now offer a basic EMF testing service for free, measuring ELF magnetic fields (power line radiation) inside and outside your home. They generally do not make recommendations for EMF reduction or assessment as to what readings are potentially harmful.

- United States Department of Energy, Bonneville Power Administration, P.O. Box 3621, Portland OR 97208-3621, Phone: (800) 622-4519 or (503) 230-3478.

  Federal government agency with informative booklets: Electrical Power Lines —Questions and Answers on Research into Health Effects and Electrical and Biological Effects of Transmission Lines — A Review.

- Check your local phone directory's yellow pages under Environmental and Ecological Services for other EMF testing services.

- International Institute for Baubiologie and Ecology, Inc., P.O. Box 387, Clearwater FL 34615, Phone: (813) 461-4371.

  Headquarters for their German Building Biology correspondence course dealing with EMF and chemical factors in the home related to health.

- TCO Information Center, 150 N. Michigan Ave., Suite 1200, Chicago IL 60601-7594, Phone: (312) 781-6223, Fax: (312) 346-0683.

  The United States affiliate of the TCO labor union in Sweden. They have informative booklets about ergonomic and EMF factors involved in computer work: Screen Checker and Screen Facts. Free information about Sweden's MPR2 and TCO computer radiation standards. United States newsletter begun in 1995; first issue had two articles regarding ES and computers.

- SWEDAC, Box 878, S-50115, Boras Sweden, Phone: +46-8 613 4002, Fax: +46-8 613 4003.

  Swedish Board for Technical Accreditation has technical publications dealing with Sweden's MPR2 standards for computer monitors.

- Check with your library for a list of magnetic shielding manufacturers in the books <u>Thomas Register of American Manufacturers</u>, Products and Services Section, Shielding: Magnetic.

- TCO, The Swedish Confederation of Professional Employees, S-11494, Stockholm, Sweden, Phone: +46 8 782 9100, Fax: + 46 8 782 92 07.

  Headquarters for the TCO labor union. Further information regarding TCO computer radiation standards available.

APPENDIX

Computer Radiation Standards

There are no mandatory United States non-ionizing radiation standards for computer monitors other than the Federal Communication Commission's (FCC) approval for radio frequency interference suppression. The Radiation Control for Health and Safety Act of 1968 required television sets, another cathode ray tube screen, to reduce ionizing radiation — primarily reducing x-ray emissions. In the 1960's some television sets were emitting significant levels of x-rays and were recalled.[96]

Unfortunately, the Radiation Control for Health and Safety Act of 1968 has not been revised to include non-ionizing radiation from electrical appliances. The standard old-fashioned phrase is that there is no established link between power line radiation or other lower frequency fields and health hazards. But the anecdotal accounts of health hazards are worldwide.[2, 12, 22, 41] A United States Department of Energy booklet reports that power line (60 Hz) radiation effected the following biological *changes* in the laboratory: changes in tumor development, white blood cell counts, animals' behavior, human heart rate, human brain activity, cell and tissue function, biorhythms, growth, etc. Additionally, animals could detect and sought to avoid strong electric (E-field) exposure.[128]

Historically, many studies show biological effects. Science debates whether these biological effects are hazardous. Some studies showed no effect to counter other studies where the effect seemed clear. Admitting health hazards from EMFs would be costly both in the face of needed changes in technology and also regarding prior liability for injuries.

So anecdotal accounts are being largely ignored by science, in favor of scientific studies of EMFs, funded by computer manufacturers, electrical utilities, other industry groups, etc. Needless to say, there is a question of potential conflict of interest in the cases where those who have the most to lose are searching for the truth about EMF health hazards.

Sweden has voluntary technology-based radiation standards for computer monitors. Their standards, called MPR2, require reduction of the primary emissions, ELF and VLF fields, based upon what is easy for industry to do at little additional cost. MPR2 is not designed to be health-based standards, because "safe" EMF levels have not been scientifically determined.[21] Sweden's white collar labor union called TCO objected to MPR2 standards, saying that they are not safe enough. Now TCO has their own standards, based upon the limits of present technology.[121] Their intention is that computer radiation be "as low as technically feasible". TCO mentions oversensitivity to electricity, skin problems, cancer and birth defects among their published concerns of computer radiation exposure based upon their group's experience.[120] (See MPR2 and TCO Standards, page 81.)

The ES patient's illness is beyond the TCO standards' helpfulness. The ES person usually cannot use a computer at all, even if shielded. Using equipment that is emitting EMFs, just weakens the ES patient's condition further in the long-term.

If you were to call the United States computer manufacturers and ask about MPR2 and TCO, the phone representatives may not know about either one. If their monitors meet MPR2, generally the representatives will know about it, and otherwise they usually do not. Even representatives at the major computer manufacturers do not generally know about TCO standards, even though Sweden's labor union now has an office in Chicago and is actively seeking all computer manufacturers to meet their radiation standards. Many United States computer manufacturers now meet Sweden's MPR2 standards and a few meet TCO standards for one computer monitor in their product line.

School districts would do well to specifically ask for MPR2 or TCO standards on all new equipment. Ideally, you would want emission measurements of all MPR2 component fields to determine by how much their monitor is below those standards. New computer monitors going into school systems are not necessarily meeting MPR2 standards; you would need to ask for it or measure for it to be sure of having lower-radiation monitors. Older monitors ideally would be replaced or upgraded to reduce computer radiation.[64]

Sweden's computer radiation guidelines are the strictest in the world. In 1991, Sweden's MPR2 guidelines took effect in Sweden, which primarily reduces the ELF and VLF frequencies from computer monitors.

ELF and VLF fields have two components — electric and magnetic fields. The electric fields are easier to reduce than the magnetic ones. Anti-radiation glare screens can significantly reduce these electric fields at the front of the monitor. Glare screen manufacturers sometimes claim that their screen blocks 98% or more of ELF/VLF radiation — but this is only the electric component (the E-Field).

The magnetic field (the H-Field) part of ELF/VLF requires a nickel/iron alloy like Mumetal to reduce these magnetic fields — the ELF magnetic field is the one currently of most concern. However, the VLF magnetic field has more energy to induce electric current and is limited by MPR2 standards also.

Glare screens with a conductive coating and grounding wire are helpful to reduce the alternating current E-field (electric field) and static electricity from the screen, but do not reduce the H-field (magnetic field) emissions. Grounded metal around the computer monitor may help reduce static and E and H field emissions, depending upon the magnetic field shielding quality of the metal. Nickel/iron alloys can offer good reduction, but at higher cost. Other metals used for magnetic shielding in general include iron, low-carbon steel, iron mixed with aluminum, silicon, and/or cobalt, etc.

Grounding refers to transferring the E-field out into the building's electrical grounding system which takes the current to the earth. Grounding is not about magnetic field reduction. Metal shielding that is highly magnetically permeable is most effective for magnetic field reduction.

Computer radiation has various levels of exposure depending upon the equipment in operation. For example, if the screen is shut off, but the CPU (central processing unit) is still on, ELF radiation emissions are occurring from the CPU. The only sure way of eliminating the computer radiation exposure is by unplugging the computer monitor and all related equipment. When a monitor is on but not in use — a common office practice — computer radiation exposure occurs. Some newer monitors have a power-down feature to limit energy use and EMFs when in the on, but inactive, mode.

However, the most serious radiation exposure occurs when the computer is actually in use. Then, there is more radiation and more intense radiation than when the computer is on but inactive. An ES patient can sometimes feel the pulses of radiation with each keystroke of the computer. A critical question is whether current Swedish standards take account of these increased fields when someone is operating the keyboard. If they currently do not, higher readings and/or more frequencies of EMFs could be accounted for by retesting for MPR2 or TCO standards with a typist using the computer to simulate actual user exposures.

Sweden's voluntary computer radiation guidelines

The TCO labor union's standards below are an attempt to bring EMF emissions to the lowest measurable level based on current technology.

	MPR2	TCO
- X-rays (measurable)	0	0
- Electrostatic potential (static electricity)	less than or equal to 500 volts	less than or equal to 500 volts
- Alternating electric field Band I (ELF) * 5 Hertz - 2 kilohertz	less than or equal to 25 volts per meter, measured at 50 centimeters (19.7 inches) in front of the computer monitor	less than or equal to 10 volts per meter, measured at 30 centimeters (11.8 inches) in front of the computer monitor.
- Alternating electric field Band II (VLF) ** 2 kilohertz - 400 kilohertz	less than or equal to 2.5 volts per meter, measured 50 centimeters (19.7 inches) around the computer monitor.	less than or equal to 1 volt per meter, measured at 50 centimeters (19.7 inches) around the computer monitor and at 30 centimeters (11.8 inches) in front of the computer monitor (single point)
- Magnetic field Band I (ELF) *5 Hertz - 2 kilohertz	less than or equal to 2.5 milligauss measured at 50 centimeters (19.7 inches) around the computer monitor	less than or equal to 2 milligauss measured to 50 centimeters (19.7 inches) around the computer monitor and at 30 centimeters (11.8 inches) in front of the computer monitor (single point)
- Magnetic field Band II (VLF)** 2 kilohertz - 400 kilohertz	less than or equal to .25 milligauss, measured at 50 centimeters (19.7 inches) around the computer monitor	less than or equal to .25 milligauss, measured at 50 centimeters (19.7 inches) around the computer monitor

*   ELF = extremely low frequency
** VLF = very low frequency
Source: TCO Information Center, 150 N. Michigan Avenue, Suite 1200, Chicago, IL
   60601-7594 Phone: (312) 781-6223[121]

Magnetic Field Readings from Common Home Appliances

(ELF Frequency Readings) *

at 12 inches from the source

	Reading
Microwave Oven	40 - 80 mG*
Clothes Washer	2 - 30 mG
Electric Range	4 - 40 mG
Electric Shaver	1 - 90 mG
Fluorescent Lamp	5 - 20 mG
Hair Dryer	1 - 70 mG
Television	.4 - 20 mG

The microwave oven, fluorescent lamp, and the television also use higher frequencies. The clothes washer, electric range, hair dryer, and electric shaver may measure readings in the very low frequency (VLF) range as well as ELF due to harmonics.

* mG = milligauss; readings above 2 mG are suspected of being related to increased cancer rates. Fields below 2 mG are detectable by many ES patients.

Source: United States Department of Energy[127]

Electric and Magnetic Field Readings
from United States High-Voltage Power Lines
(ELF Frequency Readings)

Electric Field*	115 kV	230 kV	500 kV
Maximum Reading on Right-of-Way	1.0	2.0	7.0
Distance from lines:			
50 feet (edge of Right-of-Way)	.5	1.5	-
65 feet (edge of Right-of-Way)	-	-	3.0
100 feet	.07	.3	1.0
200 feet	.01	.05	.3

Magnetic Field**:	115 kV	230 kV	500 kV
Maximum Reading on Right-of-Way			
Average	30	58	87
Peak	63	118	183
Distance from lines:			
50 feet (average/peak)	7/14	20/40	-
65 feet	-	-	30/62
100 feet			
Average	2	7	13
Peak	4	15	27
200 feet			
Average	.4	2	3
Peak	1	4	7

* in kilovolts per meter = kV/m
** in milligauss = mG. Readings above 2 mG have been related to higher cancer rates.
Source: United States Department of Energy[127]

84

Environmental Illness at the Worksite

Time and place are important determinants of environmental health symptoms. If symptoms intensify at a specific place or time during the day suspect environmental factors.

- Do symptoms abate away from the worksite (evenings, weekends, vacations)?

- Is your computer workstation ergonomically designed? Health problems at the computer need to be sorted out as to whether environmental or ergonomic factors are causing the problem — or perhaps both.

- Do symptoms occur just being at work, before work or during breaks? Indoor air quality may be poor and creating a Sick Building Syndrome problem.

Working in an office environment easily becomes impossible when a person is ES. Computer monitors, which now seem to be everywhere, are among the worst EMF offenders. Electric typewriters, laser printers, copy machines, and fax machines are other common office EMF sources.

ES people are in desperate need of assistance from the Americans with Disabilities Act, Social Security disability, Vocational Rehabilitation and other assistance programs that help the disabled, either to cope with a restructured job or be awarded disability income when no accommodations can be made and other options are not possible.

In the workplace — where much environmental illness begins — most workers are unable to change their environment without support from their employers.

If you suspect that you are having the early signs of an environmental illness from workplace chemical or electromagnetic exposures, *do not* merely try to adapt to the toxin/stressor. This quiet acceptance of the problem is how many MCS and ES problems began and developed into literally intolerable physical nightmares. Instead, speak to your supervisor about your concerns, perhaps also the personnel department or other employee grievance vehicles, such as your labor union representative. Other angles may involve the early guidance of a worker's compensation lawyer and a medical doctor knowledgeable in environmental medicine to back up your claims.

A period of leave from work may be necessary to see whether your condition improves, while the employer assesses and hopefully corrects the worksite health hazard. A trial return to work medically monitored could lead to the cause and effect connection between the workplace and your recurring symptoms for worker's compensation, Social Security disability, or other benefits.

Quitting your job to avoid the problem and trying to find another type of job while still unwell is not usually a good idea and relieves responsibility from the employer who damaged your health and reduced your ability to maintain your career and normal standard of living. However, staying at a worksite that makes you ill is unacceptable and should be reported to management. Correcting the problem, relocating you, or allowing you a leave of absence while the matter is being investigated would be appropriate to diminish further liability on the part of the employer, while not risking your health further.

Points to Consider for EMF Reduction in the Workplace

1. Measurement of worksite EMF — requires an ELF gaussmeter or an EMF testing service. (See EMF Resources, page 74.)

   - Record EMF measurements and note their location.

   - What is "background" EMF at the center of the room with equipment shut off?

   - Attempt to determine the exact source of each reading. Remember, ELF magnetic fields penetrate walls. Turning off all power at the electrical circuit box will determine whether outside power lines are producing a reading indoors.

   - Most inexpensive gaussmeters will only measure one direction of the field at a time. To find the highest reading, turn the meter horizontal, vertical, and sideways to the EMF source. A triple-axis meter will measure the three dimensions at once for one combined milligauss reading.

   - If an EMF reading is high at a wall with all appliances unplugged on both sides of the wall and outside power lines are not the contributing factor, an electrician who understands EMFs may need to be called to check building wiring.

- What are EMF readings of office equipment at operator distances?
- In areas reporting EMFs greater than 2mG, can work areas be removed to lower EMF areas?

2. Call monitor manufacturers — do workplace monitors meet MPR2 standards? If not, you could consider EMF reductions to present monitors or replacing old monitors with newer, lower-radiation MPR2 or TCO models. Reduction equipment would consist of a Mumetal type alloy for ELF and VLF magnetic field reduction, and a grounded glare screen attachment with a conductive coating to reduce ELF/VLF electric fields and static at the screen.

3. Are computer monitors spaced generously apart from one another, particularly if they do not meet MPR2.

4. Are workers spaced at a comfortable distance from their monitor? A distance of about two feet from the screen is helpful for reducing EMF exposures, but may not be comfortable for the user. Software that enlarges letter size and a keyboard drawer help increase the distance between the screen and user.

5. Clearing extra equipment (printers, fax machines, etc.) from your desk further reduces EMF exposure. Place the equipment on another table away from the immediate workstation area.

6. Turning off all computer equipment that is not actively being used reduces unnecessary EMF exposures when non-computer activities are involved.

7. If central processing units (CPUs) are sitting beneath monitors at the workplace, two EMF sources are near the user. An extension cord for the CPU allows you to place the CPU on the floor away from the worker's feet. A monitor arm can then elevate the monitor to eye level as before.

8. When shopping for new office equipment, measure the EMF first and compare

before you buy.

A Note to EMF Scientists

Why are many EMF scientific studies not successfully repeatable by other experimenters?

EMFs are very pervasive. When working in a laboratory, many confounding variables are present which can affect experimental results.

For example, fluorescent lights come in various types: full-spectrum, daylight, energy-efficient, etc. Each type of light can affect organisms and cells differently.[96]

The electromagnetic fields of fluorescent lights, computers, refrigerators, all motorized equipment, and wall wiring affect ES patients, and therefore would affect life on the microscopic level. Computer monitors and television screens are among the most EMF-penetrative sources. Computers placed within two rooms of an experiment — upstairs, downstairs, or next door — could affect study results, even if meters show no reading. Of course, no reading only means the limits of the meter have been exhausted but not necessarily the sensitivity of cells. The most sophisticated EMF detection equipment can detect computer EMFs and the information it contains from outside the building.[9, 34]

ES patients can be sensitive to computer monitors through walls (magnetic field exposure) and refrigerator motors at 15-20 feet or more. Precautions for cell studies would need to take special precautions to remove all EMF emitters not involved in the study. Stray fields from power surges in the building also need to be recorded. Magnetic field levels of the earth at that destination would ideally be noted, as they may influence test results.[40]

Essentially, the testing site must be environmentally clean of stray EMF or chemical contaminants that could confound the study. Beyond shielding the room from external EMF sources, the equipment brought into the site is also EMF emitting and needs to be minimized. For example, electron microscopes kill the sample being studied. Dark-field microscopes do not.

Air quality of the room from a chemically sensitive person's perspective could be assessed. Chemicals from the construction materials of the room and the experimenters' clothing (soap/cologne/smoke, etc.) may change the experiment.

A SELECTED BIBLIOGRAPHY

References are arranged by country, in alphabetical order. Foreign and small-release items have contact addresses, where appropriate.

Denmark

1.   El-og Billedskaermsskadede i Danmark (The electrically and VDT-injured in Denmark). Letter to author. 19 Nov. 1994.
     Contact: EBD, Lunden 1, Alum, 8900 Randers Denmark.

England

2.   Bentham, Peggy. VDU Terminal Sickness: Computer Health Risks and How to Protect Yourself. London: Green Print, 1991.
3.   Berg, Mats and Ingvar Langlet. "Defective Video Displays, Shields, and Skin Problems." The Lancet, 4 April 1987, p. 800.
4.   Choy, Ray V.S., M.B., B.S., Jean A. Monro, M.B., B.S., and Cyril W. Smith, Ph. D. "Electrical Sensitivities in Allergy Patients." Clinical Ecology, Volume IV, Number 3, pp. 93-102.
5.   Coghill, Roger. Electropollution: How to Protect Yourself Against It. Wellingborough, England: Thorsons Pub. Group, 1990
6.   Lakhovsky, Georges. The Secret of Life. Trans. Mark Clement. Surrey: True Health Publishing Co., 1951.
7.   Mansfield, Peter and Jean Monro. Chemical Children: How to Protect Your Family from Harmful Pollutants. London: Century, 1987.
8.   Smith, Cyril W., Ph.D. Letter to author. 14 Nov. 1994.
9.   Smith, Cyril W., Ph.D., and Simon Best. Electromagnetic Man: Health and Hazard in the Electrical Environment. New York: St. Martin's Press, 1989.
10. Smith, Cyril W., Ph.D., Ray Y.S. Choy, M.B., B.S., and Jean A. Monro, L.R.C.P., M.R.C.S. "The Diagnosis and Therapy of Electrical Hypersensitivities." Clinical Ecology, Volume VI, Number 4, pp. 119-128.

Germany

11. Arbeitskreis für Elektrosensible e.V. (Action Group for Electrically Sensitive). Letter to author. 4 Nov. 1994.
     Contact: Arbeitskreis für Elektrosensible e.V., Alleestr. 135, 44793 Bochum, Germany.

90

12.	Blomkvist, Anna-Christina et al. "Electric and Magnetic Sanitation of the Office." Work with Display Units 92. Ed. H. Luczak, A. Cakir, and G. Cakir. Berlin: Elsevier Science Publishers B.V., 1993, pp. 77-84. Selected Proceedings of the Third International Scientific Conference on Work with Display Units, Berlin, Germany. 1-4 Sept. 1992.

Sweden

13.	Floberg, Leif, D.Sc. The Invisible, Insidious Dangers of Magnetism. Lund, Sweden: KF - Sigma, 1993.
Contact: KF-Sigma, Sölvegatan 22, S-223 62 Lund, Sweden.

14.	Föreningen för el-och bildskärmsskadade (The Association for the Electrically and VDT Injured). Electrical Hypersensitivity. Stockholm, Sweden: FEB.
Contact: Föreningen för el-och bildskärmsskadade (FEB), Box 15126, 10465 Stockholm, Sweden or INTERNET:http://www.isy.liu.se/~tegen/febost.html or see ordering information at the back of this book.

15.	Föreningen för el-och bildskärmsskadade (The Association for the Electrically and VDT Injured). Technical Guide for the Electrically Sensitive. Stockholm, Sweden: FEB.
Contact: Föreningen för el-och bildskärmsskadade (FEB), Box 15126, 10465 Stockholm, Sweden or INTERNET:http://www.isy.liu.se/~tegen/febost.html or see ordering information at the back of this book.

16.	Fransson, K. and A. Eriksson. "A Questionnaire Study Among SIF-Members About Their Experiences of Electromagnetic Hypersensitivity." Electric Hypersensitivity. Article regarding The Swedish Union of Clerical and Technical Employees in Industry, (SIF), Stockholm, Sweden, p. D-22.
Contact: TCO Development Unit, The Swedish Confederation of Professional Employees, S-114 94, Stockholm, Sweden.

17.	Johansson, Olle and Peng-Yue Liu. ""Electrosensitivity", "Electrosupersensitivity" and "Screen Dermatitis": Preliminary Observations from On-Going Studies in the Human Skin." COST-244 Biomedical Effects of Electromagnetic Fields Workshop on Electromagnetic Hypersensitivity, Graz. 26-27 Sept. 1994.
Contact: Experimental Dermatology Unit, Department of Neuroscience, Karolinska Institute, 171 77 Stockholm, Sweden.

18.	Knave, Bengt. "Hypersensitivity to Electricity." Swedish Work Environment Fund Newsletter, 3-4 1993.
Contact: The National Institute of Occupational Health, Division of Occupational Neuromedicine, S-171 84, Solna, Sweden.

19.	Nordström, Gunni. "In a Special Taxi and an Iron-Clad Room Per Segerbäck is Forced to Live Apart from Everything Electric." TCO Newspaper, No. 18, 18 June 1993.

Contact for English Translation: Föreningen för el-och bildskärmsskadade (FEB), Box 15126, 10465 Stockholm, Sweden or INTERNET:http://www.isy.liu.se/~tegen/febost.html

20.   Södergren, Leif. "Electrical Hypersensitivity in Sweden - Uncovering the Cover Up." Article to author.
Contact: Föreningen för el-och bildskärmsskadade (FEB), Box 15126, 10465 Stockholm, Sweden or INTERNET:http://www.isy.liu.se/~tegen/febost.html

21.   SWEDAC. Test Methods for Visual Display Units: MPR 1990:8. 1990-12-01. SWEDAC, 1990.

22.   Swedish National Institute of Occupational Health. "Cancer Related to Strong Electromagnetic Fields." Forskning and Praktik, English Edition, 7/1992, pp. 3-15.
Contact: Förlagstjänst, National Institute of Occupational Health, 17184 Solna, Sweden.

23.   Swedish National Institute of Occupational Health. "One of Seven Sensitive to Electrical Fields." Forskning and Praktik, English Edition, 4/1992, pp. 3-8.
Contact: Förlagstjänst, National Institute of Occupational Health, 17184, Solna, Sweden.

USA/Other

24.   Accola, Julie. "Jet Smart." The Human Ecologist, Fall 1993, Number 59, pp. 30-31.
Contact: The Human Ecologist, PO Box 49126, Atlanta, GA 30359-1126.

25.   Aladjem, Henrietta, and Peter H. Schur, M.D. In Search of the Sun: A Woman's Courageous Victory over Lupus. New York: Scribner, 1988.

26.   American Cancer Society. Cancer Facts and Figures - 1993. Atlanta, GA: American Cancer Society, 1993.
Contact: American Cancer Society, 1599 Clifton Rd., NE, Atlanta, GA 30329-4251.

27.   American Optometric Association. VDT User's Guide to Better Vision. St. Louis, MO: American Optometric Association.
Contact: American Optometric Association, 243 N. Lindberg Blvd., St. Louis, MO 63141.

28.   American Society of Nutritional Research. "Research Bulletin #160: The Minerals." (paper). Phoenix: American Society of Nutritional Research.
Contact: American Society of Nutritional Research, PO Box 241, Phoenix, AZ 85029

29.   Arrillaga, J., D.A. Bradley, and P.S. Bodger. Power System Harmonics. Chichester (West Sussex), New York: Wiley, 1985.

92

30. Banta, John. Current Switch: How to Reduce or Eliminate Electromagnetic Pollution in the Home and Office. (video). Prescott, AZ: Healthful Hardware, 1994.
     Contact: Healthful Hardware, PO Box 3217, Prescott, AZ 86302.

31. Barnothy, M.F., ed. Biological Effects of Magnetic Fields. Vol. 1. New York: Plenum Press, 1964.

32. Barnothy, M.F., ed. Biological Effects of Magnetic Fields. Vol. 2. New York: Plenum Press, 1969.

33. Becker, Robert O., M.D., and Gary Selden. The Body Electric: Electromagnetism and the Foundation of Life. New York: Morrow, 1985.

34. Becker, Robert O., M.D. Cross Currents: The Promise of Electromedicine, the Perils of Electropollution. Los Angeles: J.P. Tarcher, 1990.

35. Beddows, Norman A., C.I.H., C.S.P. "Extremely Low Frequency (ELF) Magnetic Fields in Offices and Their Mitigation." (paper). Philadelphia: Amuneal, 1993.
     Contact: Amuneal Manufacturing Corp., 4737 Darrah St., Philadelphia, PA 19124.

36. Belt, Don. "Sweden." National Geographic, August 1993, pp. 2 - 35.

37. Bircher, Ralph, ed. Raw Foods and Juices Nutrition Plan. Trans. Timothy McManus. Los Angeles: Nash Pub., 1972.

38. Bird, Christopher. The Divining Hand: The 500-Year-Old Mystery of Dowsing. New York: Dutton, 1979.

39. Biser, Sam. "Insomnia: How to Get Major Relief in 48 Hours — Without Sleeping Pills, Herbs, or Vitamins." Advanced Natural Therapies, Volume 2, Number 8.
     Contact: The University of Natural Healing, Inc., PO Box 8113, 355 West Rio Road, Suite 201, Charlottesville, VA 22906.

40. Brodeur, Paul. Currents of Death: Power Lines, Computer Terminals, and the Attempt to Cover Up Their Threat to Your Health. New York: Simon and Schuster, 1989.

41. Brodeur, Paul. The Great Power-Line Cover-up: How the Utilities and the Government are Trying to Hide the Cancer Hazard Posed by Electromagnetic Fields. Boston: Little, Brown and Co., 1993.

42. Burr, Harold Saxton. The Fields of Life: Our Links with the Universe. New York: Ballantine Books, 1972.

43. Canby, Thomas Y. "Bacteria: Teaching Old Bugs New Tricks." National Geographic, August 1993, pp. 36-60.

44. Castleman, Michael, "Why?" Mother Jones, May/June 1994, pp. 34-42.

45. Cook, Trevor M. Homeopathic Medicine Today. New Canaan, CN: Keats Pub., 1989.

46. Crile, George, W. The Phenomena of Life: A Radio-Electric Interpretation. New York: W.W. Norton, 1936.

47. Crook, William G., M.D. <u>The Yeast Connection: A Medical Breakthrough</u>. Jackson, TN: Professional Books, 1985.

48. Crook, William G., M.D., and Marjorie Hurt Jones, R.N. <u>The Yeast Connection Cookbook</u>. Jackson, TN: Professional Books, 1989.

49. Dadd, Debra Lynn. <u>The Nontoxic Home: Protecting Yourself and Your Family From Everyday Toxics and Health Hazards</u>. Los Angeles: J.P. Tarcher, 1986.

50. Dallin, Lynn. <u>Cancer Causes and Natural Controls</u>. Port Washington, NY: Ashley Books, 1983.

51. Donsbach, Kurt W., D.C., Ph.D., and Morton Walker, D.P.M. <u>Dr. Donsbach Tells You What You Always Wanted to Know About Negative Ions</u>. Rosarito Beach, Baja, CA, Mexico: Wholistic Publications, 1981.

52. Donsbach, Kurt W., D.C., Ph.D., with Richard O. Brennan, DO. <u>Dr. Donsbach Tells You What You Always Wanted to Know About Superoxide Dismutase</u>. Rosarito Beach, Baja, CA, Mexico: Wholistic Publishers, 1988.

53. Dumpé, Bert. <u>X-Rayed Without Consent: Computer Health Hazards</u>. Arlington, VA: Ergotec Association, 1989.

54. Dutta, S.K., Ph.D., and R.M. Millis, Ph.D. eds. <u>Biological Effects of Electropollution: Brain Tumors and Experimental Models</u>. Philadelphia, PA: Information Ventures, 1986.

55. Dyer, John Robert. <u>Organic Spectral Problems</u>. Englewood Cliffs, NJ: Prentice-Hall, 1972.

56. Eisenberg, David, M.D., with Thomas Lee Wright. <u>Encounters with Qi: Exploring Chinese Medicine</u>. New York: Norton, 1985.

57. Fisher, Alexander A., M.D. " 'Terminal' Dermatitis Due to Computers (Visual Display Units)." <u>Cutis</u>, Volume 38, 1986, pp. 153-154.

58. Gazella, Karolyn A. "Attention Deficit Hyperactivity Disorder: Focusing on Alternative Treatments." <u>Health Counselor</u>, Feb. 1994, pp. 24-29. Contact: Impakt Communications, Inc., PO Box 12496, Green Bay, WI 54307-2496.

59. Gazella, Karolyn A. "Environmental Illness: Seeking Protection from a Toxic World." <u>Health Counselor</u>, Aug./Sept. 1994, pp. 26-31. Contact: Impakt Communications, Inc., PO Box 12496, Green Bay, WI 54307-2496.

60. Gerson, Charlotte. Letter to author. 5 Dec. 1994.

61. Gerson, Max M.D. <u>A Cancer Therapy: Results of Fifty Cases and the Cure of Advanced Cancer by Diet Therapy — A Summary of 30 Years of Clinical Experimentation</u>. Bonita, CA: Gerson Institute, 1990.

62. Gibilisco, Stan, ed. <u>Encyclopedia of Electronics</u>. Blue Ridge Summit, PA: TAB Professional and Reference Books, 1985.

63. Glassman, Judith. <u>The Cancer Survivors and How They Did It</u>. Garden City, NY: Dial Press; Doubleday, 1983.

64. Grant, Lucinda. <u>Workstation Radiation: How to Reduce Electromagnetic Radiation Exposure from Computers, TV Sets and Other Sources</u>. Prescott, AZ: Weldon Publishing, 1992.

65. Grover, V. Lee. "I Used to Have Electromagnetic Problems." <u>The New Reactor: The Newsletter of the Environmental Health Network</u>. Nov./Dec. 1994, p. 4.
    Contact: EHN, PO Box 1155, Larkspur CA 94977.

66. Hembree, Diana and Sarah Henry. "Hypersensitivity Stalks Silicon Valley." <u>The San Diego Union</u>, 5 Jan. 1987, p. D-4.
    Contact: The San Diego Union, 350 Camino De La Reina, PO Box 191, San Diego, CA 92112-4106.

67. Hitching, Francis. <u>Dowsing — the Psi Connection</u>. Garden City, NY: Anchor Books, 1978.

68. Hughes, Marija Matich. <u>Computer Health Hazards</u>. Volume 1. Washington, D.C.: Hughes Press, 1990.

69. Hughes, Marija Matich. <u>Computer Health Hazards</u>. Volume 2. Washington, D.C.: Hughes Press, 1993.

70. Hyman, Jane Wegscheider. <u>The Light Book: How Natural and Artificial Light Affect Our Health, Mood, and Behavior</u>. Los Angeles: Jeremy P. Tarcher, Inc., 1990.

71. The International Non-Ionizing Radiation Committee of the International Radiation Protection Association. <u>IRPA Guidelines on Protection Against Non-ionizing Radiation: The Collected Publications of the IRPA Non-ionizing Radiation Committee</u>. New York: Pergamon Press, 1991.

72. Kals, W.S. <u>Your Health, Your Moods, and the Weather</u>. Garden City, NY: Doubleday, 1982.

73. Kimball, Richard W. "Acupuncture: Most Misunderstood; Serves to Complement Traditional Forms." <u>The Prescott Courier</u>, 17 Sept. 1993.
    Contact: The Prescott Courier, PO Box 312, Prescott, AZ 86302

74. Krieger, Dolores. <u>The Therapeutic Touch: How to Use Your Hands to Help or to Heal</u>. Englewood Cliffs, NJ: Prentice-Hall, 1979.

75. Krieger, Roy W. "On the Line." <u>ABA Journal</u>, Jan. 1994, pp. 40-45.

76. Kushi, Michio. <u>The Book of Dō-In: Exercise for Physical and Spiritual Development</u>. Tokyo, Japan: Japan Publications, 1979.

77. The Labor Institute. <u>Electromagnetic Fields (EMFs) in the Modern Office: A Training Workbook for Working People</u>. New York: The Labor Institute.
    Contact: The Labor Institute, 853 Broadway, Room 2014, New York, NY 10003.

78. The Labor Institute. <u>Multiple Chemical Sensitivities at Work: A Training Workbook for Working People</u>. New York: The Labor Institute, 1993.

Contact: The Labor Institute, 853 Broadway, Room 2014, New York, NY 10003.

79.   Lamielle, Mary, ed. "Veteran's Issues." The Delicate Balance, Fall/Winter 1993-1994, pp. 21-25.

Contact: The Delicate Balance, 1100 Rural Ave., Voorhees, NJ 08043.

80.   Lawren, Bill. "The 'Juice' Cure: Why A Visit to the Doctor May Become an Electrifying Experience." Longevity, Nov. 1993, pp. 61-62, 100-102.

81.   Lindbloom, Marja-Liisa, et al. "Magnetic Fields of Video Display Terminals and Spontaneous Abortion." American Journal of Epidemiology, 1 Nov. 1992, pp. 1041-1051.

82.   "Lupus Common Diagnosis in Clinic's First Year." The Prescott Courier, 9 Jan. 1995.

83.   Lupus Foundation of American, Inc. Introduction to Lupus. Rockville, MD: Lupus Foundation of America, Inc., 1994.

Contact: Lupus Foundation of America, Inc., 4 Research Place, Suite 180, Rockville, MD 20850-3226.

84.   Lupus Foundation of America, Inc., Greater Arizona Chapter. "Health Care Reform Must Meet the Needs of People with Lupus and Other Chronic Illnesses." (paper). Phoenix: LFA.

Contact: Lupus Foundation of America, Inc., Greater Arizona Chapter, 2149 W. Indian School Rd., Phoenix, AZ 85015-4908.

85.   Lynes, Barry with John Crane. The Cancer Cure that Worked!: Fifty Years of Suppression. Toronto, Canada: Marcus Books, 1987.

86.   Matthews, Bonnye L. Chemical Sensitivity: A Guide to Coping with Hypersensitivity Syndrome, Sick Building Syndrome, and Other Environmental Illnesses. Jefferson, NC: McFarland and Co., Inc., 1992.

87.   Maugh II, Thomas H. "Alzheimer's Disease Tied to Exposure to EMFs." Des Moines Sunday Register, 31 July 1994.

88.   McGraw-Hill. Encyclopedia of Science and Technology. New York: McGraw-Hill, 1992.

89.   Morrison, Ralph. Grounding and Shielding Techniques in Instrumentation. New York: Wiley, 1986.

90.   Morse, Melvin, M.D., with Paul Perry. Transformed by the Light: The Powerful Effects of Near Death Experiences on People's Lives. New York: Villard Books, 1992.

91.   Murray, Michael T., N.D., and Joseph Pizzorno. Encyclopedia of Natural Medicine. Rocklin, CA: Prima Pub., 1991.

92.   National Association of Working Women, (9 to 5). "Facts on Fluorescents." (paper). Cleveland, OH: NAWW.

Contact: National Association of Working Women, 614 Superior Avenue, NW, Cleveland OH 44113.

93. National Association of Working Women, (9 to 5) and The Service Employees International Union. <u>VDT Syndrome: The Physical and Mental Trauma of Computer Work</u>. Cleveland, OH: NAWW, 1987.

94. National Research Council. <u>Multiple Chemical Sensitivities: Addendum to Biologic Markers in Immunotoxicology</u>. Washington, DC: National Academy Press, 1992.

95. Ott, John N. "Color and Light: Their Effects on Plants, Animals and People." <u>The International Journal of Biosocial Research</u>, Volume 7, 1985.
   Contact: The International Journal of Biosocial Research, PO Box 1174, Tacoma, WA 98401.

96. Ott, John N. <u>Health and Light: The Effects of Natural and Artificial Light on Man and Other Living Things</u>. Old Greenwich, CT: Devin-Adair Co., 1973.

97. Ott, John N. <u>Light, Radiation and You: How To Stay Healthy</u>. Old Greenwich, CT: Devin - Adair Co., 1990.

98. Paddock, Richard C. "Hypoallergenic Housing Project a Disappointment." <u>Los Angeles Times</u>, 24 Dec. 1994.

99. Persinger, Michael A., ed. <u>ELF and VLF Electromagnetic Field Effects</u>. New York: Plenum Press, 1974.

100. Poch, David I. <u>Radiation Alert: A Consumer's Guide to Radiation</u>. Garden City, NY: Doubleday, 1985.

101. Rea, William J., M.D., F.A.C.S. Letter to author. 5 Nov. 1994.

102. Rea, William J., M.D., F.A.C.S, et al. "Electromagnetic Field Sensitivity." <u>Journal of Bioelectricity</u>, Volume 10 (1 and 2), 1991, pp. 241-256.

103. Richardson, Sarah. <u>Homeopathy: Stimulating the Body's Natural Immune System</u>. New York: Harmony Books, 1988.

104. Riley, Karl. "The Professional EMF Home Survey." <u>EMF Health Report</u>, Volume 2, Number 4, 1994.
   Contact: Information Ventures, Inc., INTERNET:kleinste@eniac.seas.upenn.edu

105. Rochlitz, Steven. <u>Allergies and Candida: With The Physicist's Rapid Solution</u>. Mahopac, NY: Human Ecology Balancing Sciences, Inc., 1991.

106. Rogers, Sherry A., M.D. <u>Tired or Toxic? A Blueprint for Health</u>. Syracuse, NY: Prestige Publishers, 1990.

107. Rothschild, Peter, Ph.D., M.S., N.D., "Self-Help for Internal Cleansing." (paper).
   Contact: American Academy of Naturopathic Medicine, Inc., 8021 L. St. NW, Suite 250, Washington DC 20036.

108. Schechter, Steven R., with Tom Monte. <u>Fighting Radiation with Food, Herbs, and Vitamins: Documented Natural Remedies that Protect You from Radiation, X-Rays, and Chemical Pollutants</u>. Brookline, MA: East West Health Books, 1988.

109. Schultz, Ruth. "Electrical Sensitivity: A Highly Charged Issue." <u>Network News</u>, Oct./Nov. 1994, pp. 1-2.

Contact: Network News, 410 W 53rd St., Suite 402, New York, NY 10019.

110. Schultz, Ruth. "Electrical Sensitivity in Wisconsin: Part II of a Series." Network News, Dec. 1994/Jan. 1995, pp. 10-11.

Contact: Network News, 410 W 53rd St., Suite 402, New York, NY 10019.

111. Scott, Jimmy, with Kathleen Goss. Cure Your Allergies in Minutes. San Francisco: Health Kinesiology Publications, 1988.

112. Slesin, Louis, Ph.D., ed. "German Animal Studies Support EMF-Breast Cancer Link: A Boost for the Melatonin Hypothesis." Microwave News, Volume 13, Number 4, pp. 1-2.

Contact: Microwave News, PO Box 1799, Grand Central Station, NY 10163.

113. Slesin, Louis, Ph.D., ed. "St. Louis Brain Tumor Cluster Unresolved." VDT News, Volume 11, Number 3, pp. 6-7.

Contact; VDT News, PO Box 1799, Grand Central Station, NY 10163.

114. Slonim, N. Balfour, M.D, Ph.D., ed. Environmental Physiology. St. Louis, MO: C.V. Mosby Co., 1974.

115. Soyka, Fred. The Ion Effect: How Air Electricity Rules Your Life and Health. New York: Dutton, 1977.

116. "Supposed Ecologically Safe Home Makes the Chemically Sensitive Ill." The Prescott Courier, 11 Dec., 1994.

Contact: The Prescott Courier, PO Box 312, Prescott , AZ 86302.

117. Sugarman, Ellen. Warning — The Electricity Around You May Be Hazardous To Your Health." New York: Simon and Schuster, 1992.

118. Swoboda, Frank. "Repeat -Stress Injuries Up Sharply in OSHA Report." Los Angeles Times, 12 Dec. 1994.

119. Szent-Gyorgyi, Albert. Introduction to Sub-Molecular Biology. Academic Press, 1960.

120. TCO, The Swedish Confederation of Professional Employees. "Environmental Labelling of Display Units. Chicago: TCO.

Contact: TCO Information Center, 150 N. Michigan Ave., Suite 1200, Chicago, IL 60601-7594.

121. TCO, The Swedish Confederation of Professional Employees. "MPR 2 and TCO Guidelines." (paper). Chicago: TCO.

Contact: TCO Information Center, 150 N. Michigan Ave., Suite 1200, Chicago, IL 60601-7594.

122. TCO, The Swedish Confederation of Professional Employees. Screen Checker. English Edition. Stockholm, Sweden: TCO, 1992.

Contact: TCO Information Center, 150 N. Michigan Avenue, Suite 1200, Chicago, IL 60601-7594.

98

123. TCO, The Swedish Confederation of Professional Employees. <u>Screen Facts</u>. English Edition. Stockholm, Sweden: TCO, 1991.
  Contact: TCO Information Center, 150 N. Michigan Ave., Suite 1200, Chicago, IL 60601-7594.

124. Tips, Jack, Ph.D. <u>Your Liver...Your Lifeline: Insights on Health, Based on the Liver Triad of A. Stuart Wheelwright</u>. Austin, TX: Apple-a-day Press, 1993.

125. U.S. Atomic Energy Commission. Division of Technical Information. <u>Spectroscopy</u>. Oak Ridge, TN: GPO, 1968.

126. U.S. Congress. Office of Technology Assessment. <u>Biological Effects of Power Frequency Electric and Magnetic Fields</u>. Washington D.C.: GPO, 1989.

127. U.S. Department of Energy. Bonneville Power Administration. <u>Electrical and Biological Effects of Transmission Lines: A Review</u>. Portland, OR: DOE, 1993.
  Contact: Bonneville Power Administration, PO Box 3621, Portland, OR 97208-3621.

128. U.S. Department of Energy. Bonneville Power Administration. <u>Electric Power Lines: Questions and Answers on Research into Health Effects</u>. Portland, OR: DOE, 1994.
  Contact: Bonneville Power Administration, PO Box 3621, Portland, OR 97208-3621.

129. U. S. General Accounting Office. <u>Electromagnetic Fields: Federal Efforts to Determine Health Effects Are Behind Schedule</u>. Washington D.C. : GPO, 1994.

130. U.S. National Institute on Drug Abuse. <u>Assessing Neurotoxicity of Drugs of Abuse</u>. Rockville, MD: National Institute on Drug Abuse, 1993.

131. U.S. National Institutes of Health, National Cancer Institute. "Lifetime Probability of Breast Cancer in American Women." (paper). Bethesda, MD: NCI, 1992.
  Contact: National Cancer Institute, Office of Cancer Communications, Building 31, Room 10A24, Bethesda, MD 20892.

132. Vidali, Gianfranco. <u>Superconductivity: The Next Revolution?</u> New York: Cambridge University Press, 1993.

133. Vithoulkas, George. <u>Homeopathy: Medicine of the New Man</u>. New York: Prentice Hall, 1987.

134. Wallach, Charles Ph.D. Letter to author. 20 Nov. 1994.

135. Wilson, Bary W. "Chronic Exposure to ELF Fields May Induce Depression." <u>Bioelectromagnetics,</u> Volume 9, 1988, pp. 195-205.

136. World Health Organization. <u>Electromagnetic Fields 300 Hz to 300 GHz</u>. Geneva, Switzerland: WHO, 1993.

INDEX

Electromagnetic field (EMF) information resources available from Weldon Publishing:

- **WORKSTATION RADIATION: How to Reduce Electromagnetic Radiation Exposure from Computers, TV Sets, and Other Sources by Lucinda Grant (ISBN 0-9635407-1-8)**

  "... a practical guide..."
  The National Center for Environmental Health Strategies

  Price: $9.45

- **Electrical Hypersensitivity by The Association for the Electrically and VDT Injured - Sweden**

  The latest health information about illness related to electricity: what electrical hypersensitivity is, technical guide for coping with this disabling illness, and support groups worldwide.

  Price: $5.00 Unbound
  $8.00 Library Binding (plastic)

- **The Electrical Sensitivity Handbook: How Electromagnetic Fields (EMFs) are Making People Sick by Lucinda Grant (ISBN 0-9635407-2-6)**

  A resource guide for the latest environmental illness — electrical sensitivity from electromagnetic fields (EMFs). This international exposé explains the intricacies of the disease with resources for medical, legal, and EMF services in the United States. Includes bibliography of related EMF literature.

  Price: $23.00 USA
  $28.00 Foreign

Weldon Publishing
P. O. Box 4146
Prescott, AZ 86302